Table of Contents

Understanding Signal Processing A Comprehensive Guide to Learn Signal Processing

2

Cybellium Ltd

1. Introduction to Signal Processing

Signal processing is an intricate field at the intersection of mathematics, engineering, and computer science, dedicated to the analysis, transformation, and synthesis of signals. These signals, which can be audio, video, sensor data, or more abstract representations, underpin much of modern technology, from everyday devices to sophisticated scientific instruments. The fundamental challenge lies in manipulating these signals to extract meaningful information, enhance quality, or transform them into a more desirable form.

To grasp the vast landscape of signal processing, it is crucial to understand its foundational concepts, tools, and techniques. This chapter embarks on a journey to decode the essentials, laying the groundwork for more advanced topics. We begin by defining what a signal is and elucidating the kinds of signals commonly encountered. We will also explore the historical context and evolution of signal processing, shedding light on its rapid expansion and its role in driving technological advancements.

From telecommunications to medical imaging, from audio engineering to radar systems, signal processing is an underpinning technology. As we delve into this chapter, we will uncover the core principles that allow us to systematically approach problems involving signals, introduce the terminology and key mathematical constructs, and provide a primer on the methodologies and applications that make signal processing such a vital and dynamic field of study. Through these foundational insights, we will be prepared to navigate the more complex and nuanced aspects of the discipline in subsequent chapters.

1.1 Definition and Scope of Signal Processing

Signal processing is a critical discipline within electrical engineering and applied mathematics focused on the analysis, modification, and synthesis of signals. Signals, in this context, are defined as functions that convey information about the behavior or attributes of some phenomenon. These could be anything from audio signals, which represent sound waves, to electromagnetic signals used in communication systems, biological signals such as those recorded in medical diagnostics, or even financial signals representing stock market data. The breadth and versatility of signal processing underline its extensive range of applications and notably broad scope.

At its core, signal processing involves four primary functions: acquisition, transformation, analysis, and interpretation of signals. The acquisition refers to capturing the physical or digital representation of a signal, typically through the use of sensors or transducers. Once captured, signals often require transformation to align with the needs of a specific application. This transformation might involve filtering out noise, compressing data for efficient storage and transmission, or converting signals from one domain to another, such as from time to frequency domain via Fourier Transform.

Another vital aspect of signal processing is the analysis phase, where the properties and behaviour of the signals are studied. This can include examining the signal's frequency content, amplitude, phase, and other attributes to identify patterns, anomalies, or to extract relevant information. Signal analysis tools are diverse, ranging from basic statistical methods to sophisticated algorithms such as wavelet transforms or machine learning techniques. Finally, the interpretation of signals involves making sense of the processed data, often guiding decision-making in real-time applications like radar systems, medical imaging, or automated trading systems.

Beyond these core functions, signal processing encompasses a plethora of subfields, each targeting specific aspects and applications. Digital signal processing (DSP) and analog signal processing are two fundamental branches. DSP involves handling signals in a digital form, and thus applies discrete mathematics and algorithms. Analog signal processing, on the other hand, deals with signals in their continuous form, often utilizing analog electronic circuits.

Digital signal processing, in particular, has seen exponential growth with advancements in computer technology. Techniques such as convolution, filtering, sampling, quantization, and the implementation of transforms like the Discrete Fourier Transform (DFT) and its faster counterpart, the Fast Fourier Transform (FFT), are fundamental to DSP. These techniques underpin critical technologies in telecommunications, audio and speech processing, image and video compression, and many more fields. For instance, mobile phones rely on DSP for managing signal reception and transmission, smartphones use it for image enhancement and voice recognition, and streaming services depend on it to deliver high-quality audio and video.

In the realm of analog signal processing, essential techniques include modulation, demodulation, and the implementation of analog filters. These are integral in radio broadcasting, television transmission, and other forms of analog communication. Despite the increasing dominance of digital methods, analog processing remains pertinent in areas where digital equivalents may not be feasible or required, often at the front end of signal acquisition systems where initial signal conditions are inherently analog.

The scope of signal processing also extends beyond traditional engineering and into interdisciplinary fields such as biomedical engineering, neurosciences, and financial engineering. In biomedical engineering, for example, signal processing techniques are crucial for interpreting the electrical activity of the heart and brain, as seen in ECGs (electrocardiograms) and EEGs (electroencephalograms). Neural signal analysis assists in understanding brain functions and developing technologies like brain-computer interfaces. Financial signal processing applies these techniques to analyze stock prices, detect market trends, and manage investment risks.

Furthermore, signal processing is evolving with the advent of data science and artificial intelligence. Machine learning algorithms are increasingly being integrated with traditional signal processing techniques to enhance performance in various applications. For instance, machine learning aids in speech recognition technology, enabling more accurate and context-aware systems. In medical imaging, AI algorithms improve diagnostic accuracy by better interpreting MRI and CT scans.

In conclusion, the definition and scope of signal processing are vast and continually expanding. It encompasses the essential functions of acquiring, transforming, analyzing, and interpreting signals across multiple domains and applications. Whether in telecommunications, medical diagnostics, or financial markets, the principles of signal processing are integral to interpreting and managing information efficiently and effectively. With ongoing advancements in technology, the potential applications and methodologies within signal processing are bound to grow, presenting new challenges and opportunities in this ever-evolving field.

1.2 The Role of Signal Processing in Modern Technology

Signal processing has emerged as a foundational aspect of modern technology, underpinning an array of applications across various fields such as telecommunications, healthcare, finance, entertainment, and more. The profound impact of signal processing on these industries is attributed to its ability to manipulate, analyze, and transform information-bearing signals to enhance their quality, extract meaningful data, and ensure efficient transmission and storage.

One of the primary roles of signal processing is in telecommunications, where it has revolutionized the way we communicate. In this domain, signal processing is crucial for tasks such as data compression, error detection and correction, modulation, and filtering. Compression algorithms, like those used in video streaming and online conferencing, allow for the efficient use of bandwidth by reducing the amount of data that needs to be transmitted without significantly compromising quality. Error detection and correction mechanisms ensure that data corrupted by noise or interference during transmission is identified and rectified, thus maintaining the integrity of communication systems.

In the medical field, signal processing is indispensable for diagnostic and therapeutic technologies. Electrocardiograms (ECGs) and electroencephalograms (EEGs) are prime examples where signal processing techniques are employed to analyze heartbeats and brain activity, respectively. The analysis of these signals can detect abnormalities and aid in diagnoses, ultimately contributing to improved patient outcomes. Furthermore, imaging techniques like MRI and CT scans rely heavily on sophisticated signal processing algorithms to reconstruct high-resolution images from raw data, thereby enabling precise visualization of internal body structures.

The financial sector benefits from signal processing through techniques that analyze stock market trends, forecast financial indices, and manage risks. High-frequency trading, for instance, depends on the ability to process large volumes of data quickly and accurately to make informed trading decisions within microseconds. Signal processing algorithms are used to identify patterns and anomalies in financial data, facilitating better investment strategies and risk assessments.

In the realm of entertainment, signal processing has transformed audio and visual experiences. Audio signal processing techniques, such as noise reduction, equalization, and compression, are used in music production, broadcast, and telephony to enhance sound quality. In the film and television industry, video signal processing is essential for editing, special effects, and broadcasting, ensuring high-quality visual output. Innovations like virtual and augmented reality also rely significantly on signal processing to create immersive and responsive environments.

Beyond these applications, signal processing plays a critical role in the development and functioning of sensor networks. These networks are integral to various modern technologies, including environmental monitoring, smart grids, and the Internet of Things (IoT). By analyzing data collected from sensors, signal processing algorithms can detect changes, predict trends, and trigger appropriate responses, thereby enabling automated and intelligent systems.

Moreover, advancements in machine learning and artificial intelligence (AI) owe much to signal processing. Techniques such as feature extraction, dimensionality reduction, and noise filtering are crucial for improving the performance of machine learning models. Signal processing helps in preprocessing data to ensure that it is clean, relevant, and appropriately represented for algorithms to learn from. This synergy between signal processing and AI continues to drive innovations in fields like autonomous vehicles, speech recognition, and robotics.

In conclusion, the role of signal processing in modern technology is both expansive and indispensable. It serves as the backbone for enhancing and ensuring the reliability of communication systems, medical diagnostics, financial analytics, entertainment media, sensor networks, and AI applications. As technology progresses, the importance of signal processing will only continue to grow, enabling further advancements and efficiencies across diverse industries. Its contributions to improving the quality, accuracy, and efficiency of data handling and interpretation underscore its critical significance in the technological landscape.

1.3 Historical Perspectives on Signal Processing

Signal processing, a domain within electrical engineering and applied mathematics, fundamentally deals with the analysis, modification, and synthesis of signals. It has evolved significantly through various historical perspectives, shaped by technological advancements and theoretical breakthroughs. Tracing the evolution of signal processing is akin to exploring the broader trajectory of scientific and engineering progress.

The roots of signal processing can be traced back to ancient civilizations where early forms of signal interpretation were essential for communication. For instance, early humans used drums, smoke signals, and other rudimentary methods to convey information across distances. These primitive techniques signify early attempts at manipulating signals to facilitate communication.

Fast forward to the 19th century, the development of telegraphy marked a significant milestone as it represented one of the first instances of electronic signal processing. Samuel Morse's invention of the telegraph in 1837 and the subsequent development of Morse code provided a means for encoding and transmission of messages over long distances. This era laid the groundwork for modern signal processing by highlighting the importance of encoding signals to reduce noise and improve clarity.

The turn of the 20th century ushered in the era of telephony, further advancing signal processing. Alexander Graham Bell's invention of the telephone in 1876 introduced continuous, or analog, signal transmission for voice communication. The telephone network became a massive driving force for the development of signal processing techniques to enhance the quality of voice transmission, reduce noise, and improve reliability. Concepts like amplitude modulation (AM) and frequency modulation (FM) were introduced, laying the groundwork for more sophisticated signal processing methods.

The advent of World War II necessitated rapid advancements in signal processing technologies. Radar and sonar, essential for military operations, required innovative approaches for detecting and interpreting signals amidst the noise. Claude Shannon's groundbreaking 1948 paper, "A Mathematical Theory of Communication," revolutionized signal processing by introducing the concept of information theory. Shannon quantified information and introduced principles such as entropy, channel capacity, and redundancy. His work provided the mathematical foundation for digital signal processing (DSP), a field that would soon transform industries.

Post-World War II, the transition from analog to digital signal processing began to gain momentum. The invention of the transistor in 1947 and the development of integrated circuits in the 1960s facilitated the miniaturization of electronic components and the increased computational capabilities necessary for digital processing. Digital techniques allowed for more efficient and accurate manipulation of signals, paving the way for advancements in various applications including telecommunications, audio processing, and image analysis.

The 1970s and 1980s witnessed significant strides in digital signal processing with the introduction of digital signal processors (DSPs) and the Fourier Transform, particularly the Fast Fourier Transform (FFT) algorithm developed by Cooley and Tukey. The FFT made it computationally feasible to transform signals between the time and frequency domains efficiently, revolutionizing numerous applications in engineering and science. Additionally, digital filtering techniques, error correction codes, and compression algorithms became indispensable tools in the signal processing arsenal.

The rise of personal computing in the 1990s and the subsequent proliferation of digital devices heralded the modern era of signal processing. Consumer electronics, such as digital cameras, mobile phones, and multimedia players, relied heavily on advanced signal processing techniques for enhanced performance and functionality. The internet and the World Wide Web also spurred the development of new signal processing algorithms for data compression, streaming, and encryption.

Today, signal processing continues to evolve, driven by emerging technologies such as machine learning, artificial intelligence, and big data. Modern applications include medical imaging, speech recognition, autonomous vehicles, and the Internet of Things (IoT). Techniques like deep learning are being integrated with traditional signal processing methods to achieve unprecedented levels of sophistication and accuracy.

In summary, the historical perspectives on signal processing reveal a field that has grown from simple communication methods to a complex and integral part of modern technology. From ancient drums and telegraphs to digital computers and AI-driven algorithms, the journey of signal processing reflects the broader narrative of human ingenuity and technological progress.

1.4 Key Concepts in Signal Processing

Signal processing is a multidisciplinary field concerned with the analysis, manipulation, and interpretation of signals. Signals can be electrical, acoustic, or optical, among other types, and they carry information that can be analyzed for various purposes. Understanding the key concepts in signal processing is essential for designing systems that can filter, compress, and interpret this information efficiently. Several fundamental principles underpin the field of signal processing, and gaining a deep understanding of these principles is crucial for anyone working in this area.

One of the primary concepts in signal processing is the distinction between analog and digital signals. Analog signals are continuous and can take any value within a range, while digital signals are discrete, representing information using a sequence of numbers, typically in binary form. Converting an analog signal to a digital one involves sampling, where the continuous signal is measured at discrete time intervals. The Nyquist-Shannon sampling theorem is a cornerstone in this area, stating that a continuous signal can be perfectly reconstructed from its samples if the sampling rate is at least twice the highest frequency present in the signal. This concept is crucial for ensuring accurate digital representations of real-world signals.

Another key concept is the Fourier transform, which decomposes a signal into its constituent frequencies. The Fourier transform provides a frequency-domain representation of a time-domain signal, making it easier to analyze and manipulate certain aspects of the signal. For instance, noise removal and signal filtering are more straightforward in the frequency domain. The Discrete Fourier Transform (DFT), an essential tool in digital signal processing, allows for frequency analysis of discrete signals and is efficiently computed using the Fast Fourier Transform (FFT) algorithm.

Filtering is a central operation in signal processing, involving the alteration of a signal to remove unwanted components or enhance specific features. Filters can be classified into several types, including low-pass, high-pass, band-pass, and band-stop filters, each serving a specific purpose. For example, a low-pass filter allows frequencies below a certain cutoff frequency to pass through while attenuating higher frequencies. Digital filters can be implemented using various algorithms, with Infinite Impulse Response (IIR) and Finite Impulse Response (FIR) being the two primary types. IIR filters are recursive and use feedback, whereas FIR filters are non-recursive and rely solely on current and past input values.

Signal compression is another fundamental area within signal processing, aimed at reducing the amount of data required to represent a signal. This is particularly important for storage and transmission efficiency. Compression can be lossless, where the original signal is perfectly reconstructed, or lossy, where some information is lost but the result is still an acceptable approximation of the original signal. Techniques such as Huffman coding, run-length encoding, and transform coding (e.g., JPEG for images, MP3 for audio) are commonly used for this purpose.

Modulation is a technique used to embed information within a carrier signal to facilitate transmission over communication channels. Various modulation schemes, such as amplitude modulation (AM), frequency modulation (FM), and phase modulation (PM), alter different aspects of the carrier signal to encode the information. Understanding these techniques is vital for designing communication systems that can transmit data reliably and efficiently over varying channel conditions.

Finally, adaptive signal processing involves systems that can adjust their parameters in real-time based on the characteristics of the received signal. This adaptability is particularly useful in environments where signal conditions change dynamically, such as in mobile communication systems. Algorithms like the Least Mean Squares (LMS) and Recursive Least Squares (RLS) are commonly used for adaptive filtering, allowing systems to converge to the optimal filter coefficients iteratively.

In summary, the domain of signal processing encompasses a broad array of concepts and techniques essential for interpreting and manipulating signals. From the basic understanding of analog and digital signals to the intricacies of Fourier transforms, filtering, compression, modulation, and adaptive processing, each concept plays a critical role in enabling the practical application of signal processing in various fields such as telecommunications, audio and video technology, and medical imaging.

Mastery of these key concepts provides the foundational knowledge required to innovate and excel in the rapidly evolving landscape of signal processing.

2. The Signal Processing Ecosystem

In the contemporary landscape of technology, signal processing stands as a cornerstone, influencing myriad aspects of our daily lives and the broader scientific community alike. From the haze of sound waves forming coherent speech in telecommunication systems to the intricate dance of electrical impulses captured in medical diagnostics, the applications of signal processing are vast and relentless in their significance. This chapter delves into the expansive ecosystem where signal processing thrives, exploring the diverse array of tools, methods, and contexts in which it is employed.

Signal processing is not an isolated discipline; it is a vibrant domain that intersects with computer science, electrical engineering, statistics, and even the life sciences. As we peel back the layers of this field, we will uncover the foundational principles that drive signal processing techniques and observe how these principles are applied in both straightforward and highly complex scenarios.

In an era characterized by rapid technological advancements, understanding the ecosystem of signal processing equips us with the knowledge to innovate and solve real-world problems. This journey will encompass the intricate algorithms used to filter and transform signals, the hardware that makes signal manipulation possible, and the emerging trends poised to redefine the future of signal processing. As we embark on this exploration, let us first delineate the core components of the signal processing ecosystem and set the stage for a deeper dive into its multifaceted applications.

2.1 Components of Effective Signal Processing

Signal processing is a field that encompasses the analysis, modification, and synthesis of signals such as sound, images, and scientific measurements. Effective signal processing is pivotal in a range of industries, from telecommunications to neuroscience, enabling the extraction of valuable information from raw data. To achieve this, several core components must operate harmoniously. These components include analog-to-digital conversion, filtering, modulation, and the application of algorithms for signal analysis. Understanding each of these elements allows for the development of robust and versatile signal processing systems.

One of the foundational steps in signal processing is the analog-to-digital conversion. Signals in the real world are typically analog, meaning they have continuous values. However, digital systems, which are extensively used today, require signals to be in discrete form. Analog-to-digital conversion involves sampling the continuous signal at specific intervals (sampling rate) and quantizing its amplitude into discrete levels. This conversion is vitally important as it enables the manipulation of signals using digital processors. The accuracy of this process depends on the Nyquist theorem, which states that the sampling rate must be at least twice the highest frequency present in the signal to avoid aliasing, a phenomenon where high-frequency components are misrepresented as lower frequencies.

Once the signal is in a digital form, filtering is often the next critical step. Filtering is used to remove unwanted components from a signal, such as noise or specific frequency bands. Filters can be designed as low-pass, high-pass, band-pass, or band-stop, each serving a unique purpose. For instance, a low-pass filter allows signals with a frequency lower than a certain cutoff frequency to pass through, attenuating the higher frequency components. Digital filters can be implemented using different algorithms, such as Finite Impulse Response (FIR) and Infinite Impulse Response (IIR). Each method has its advantages and trade-offs in terms of complexity, stability, and performance.

Modulation is another vital component. It involves varying a carrier signal in order to transmit information. In wireless communications, for example, modulation schemes such as Amplitude Modulation (AM), Frequency Modulation (FM), and Phase Modulation (PM) are commonly used. Modulation converts the baseband data into a form that can travel over a medium—be it a copper wire, fiber optic cable, or airwaves—efficiently and with minimal loss. The choice of modulation technique impacts the robustness, bandwidth efficiency, and power requirements of the entire communication system.

Signal analysis algorithms are employed to extract meaningful information from the processed signals. These algorithms can range from simple statistical methods to complex machine learning models. Time-domain analysis might involve computing metrics like mean, variance, or signal energy, while frequency-domain analysis could involve Fast Fourier Transform (FFT) to examine the signal's spectral content. Advanced techniques, such as wavelet transforms, allow multi-resolution analysis, providing insights at various scales. In the realm of machine learning, algorithms like Principal Component Analysis (PCA) and neural networks help in pattern recognition and anomaly detection within the signal data.

Incorporating error detection and correction methods is also another integral component of effective signal processing. These methods ensure the reliability and integrity of the signal, especially in noisy environments or over long transmission distances. Techniques such as parity checks, cyclic redundancy checks (CRC), and more sophisticated algorithms like Reed-Solomon and Turbo codes are often employed to identify and correct errors, ensuring that the received signal accurately represents the transmitted data.

The orchestration of these components—analog-to-digital conversion, filtering, modulation, signal analysis, and error correction—forms the backbone of effective signal processing. Each component serves a specific role but is interdependent on the others to ensure the fidelity and utility of the processed signal. Understanding the principles and nuances of these elements can lead to more efficient and innovative applications, driving advancements in fields as diverse as medical imaging, audio engineering, and telecommunications.

2.2 The Role of Algorithms and Tools

Algorithms and tools have become indispensable in the modern world, permeating nearly every sector from healthcare to finance, manufacturing to entertainment. Their role in facilitating complex decision-making processes, optimizing operational efficiencies, and providing accurate data analytics cannot be overstated.

In the realm of healthcare, algorithms play a crucial role in diagnostics, treatment plans, and patient care management. For instance, machine learning algorithms can analyze vast amounts of medical data to improve diagnostic accuracy. Tools like IBM's Watson Health provide healthcare professionals with insights that might be missed in manual reviews. Algorithms can predict patient outcomes based on historical data, allowing for timely intervention and personalized treatments. Beyond diagnosis, algorithms help in optimizing hospital operations by managing schedules, resources, and patient flow more efficiently.

The financial sector has also seen transformative impacts thanks to algorithms and tools. Algorithms drive high-frequency trading, where trades are executed in fractions of a second based on real-time data analysis. Risk management tools evaluate market trends and forecast potential pitfalls, enabling financial institutions to make informed decisions. Banking algorithms assess credit scores, detect fraudulent activity, and automate customer service inquiries, thereby enhancing accuracy and operational efficiency. Robo-advisors use algorithms to manage investment portfolios, offering personalized advice at a lower cost compared to traditional financial advisors.

In the manufacturing industry, algorithms optimize supply chain logistics, manage inventory levels, and predict maintenance needs for machinery. Predictive maintenance tools use algorithms to analyze data from sensors embedded in machinery, forecasting when equipment might fail and scheduling maintenance before a breakdown occurs. This proactive approach reduces downtime and operational costs. Additionally, algorithms facilitate quality control by identifying defects in products faster and more accurately than human inspectors.

In the world of entertainment, recommendation algorithms shape the user experience on platforms like Netflix, Spotify, and YouTube. These algorithms analyze user behavior and preferences to suggest content that aligns with individual tastes, thereby enhancing user engagement and satisfaction. Tools for content creation, such as Adobe Creative Cloud, employ algorithms to streamline tasks like video editing, graphic design, and animation. Algorithms also power special effects in movies, enabling filmmakers to create visually stunning experiences.

Algorithms and tools are equally transformative in the realm of data analysis. Tools such as Microsoft Excel, Tableau, and SQL provide the means to sift through vast data sets, identify patterns, and generate actionable insights. In business contexts, these tools inform strategies on marketing, customer engagement, and product development. For example, customer relationship management (CRM) systems use algorithms to analyze customer data, segment audiences, and personalize marketing campaigns. In research, algorithms assist in processing large-scale surveys and experiments, providing clarity and direction for future studies.

In the educational sector, algorithms personalize learning experiences, adapting to the individual needs of each student. Tools like AI-powered tutoring systems provide tailored learning paths, identify knowledge gaps, and offer corrective feedback in real-time. This individualized approach helps students to learn at their own pace, increasing both engagement and retention. Furthermore, algorithms facilitate administrative tasks such as grading, scheduling, and even predicting student outcomes based on historical data.

In cybersecurity, algorithms are pivotal in identifying and mitigating threats. Machine learning models analyze patterns of network traffic and user behavior to detect anomalies that may indicate a security breach. Tools used in cybersecurity deploy algorithms for intrusion detection, malware analysis, and threat intelligence. The rapid identification of threats and automated responses enabled by these algorithms protect sensitive information and maintain the integrity of systems.

As technological advancements continue to evolve, the role of algorithms and tools will only expand, offering ever more refined solutions to contemporary challenges. While the benefits are vast, it is also essential to address the ethical considerations and potential biases inherent in algorithmic decision-making. Ensuring transparency, fairness, and accountability in the development and deployment of these algorithms will be critical in maximizing their positive impact on society.

In conclusion, the role of algorithms and tools spans multiple industries and applications, improving efficiency, accuracy, and decision-making processes across various domains. Their contributions are not only enhancing operational functions but also paving new pathways for innovation and growth, making them indispensable in the modern age.

2.3 The Influence of Technology on Signal Processing

Over the past few decades, the rapid advancement of technology has profoundly influenced the field of signal processing. Signal processing—the analysis, interpretation, and manipulation of signals—benefits from technological growth in myriad ways, transforming communications, medical diagnostics, audio and video recording, and more. This section explores the technological contributions that have reshaped this critical discipline.

Initially, signal processing relied heavily on analog methods. Early techniques were laborious and often limited by the physical properties of the components involved. However, the transition from analog to digital signal processing (DSP) marked a significant leap forward. Digital technology allowed for greater flexibility, precision, and efficiency. This shift was largely driven by advancements in microprocessors and microcontrollers, which enabled the implementation of complex algorithms that were previously impractical with analog circuitry.

The advent of powerful computing systems revolutionized signal processing by providing the necessary computational resources. Moore's Law, the observation that the number of transistors on microchips doubles approximately every two years, aptly describes the exponential growth of computing power. This growth has enabled real-time processing of complex signals, something that was impossible with earlier technology. High-performance processors, multi-core architectures, and specialized hardware accelerators such as Graphics Processing Units (GPUs) and Field Programmable Gate Arrays (FPGAs) have further enhanced the capabilities of signal processing systems.

Software development has also played a crucial role in this evolution. Advances in algorithms and the development of sophisticated software tools have expanded the horizons of signal processing. Machine learning and artificial intelligence have introduced new dimensions to the field, enabling systems to learn from data and improve over time. These technologies are particularly beneficial in pattern recognition tasks, such as speech and image recognition, where traditional methods fall short. Modern signal processing applications often leverage machine learning models to achieve high accuracy and efficiency.

Networking technologies have similarly influenced signal processing. The proliferation of high-speed internet and wireless communication standards like 4G and 5G has necessitated the development of advanced signal processing techniques to handle the increased data throughput and improve signal quality. In telecommunications, innovations such as Multiple Input Multiple Output (MIMO) systems and Orthogonal Frequency Division Multiplexing (OFDM) are direct results of advancements in digital and wireless communication technology. These techniques optimize bandwidth usage and minimize interference, thus enhancing overall communication performance.

Medical signal processing is another area where technology has made profound impacts. Techniques such as Magnetic Resonance Imaging (MRI), Electroencephalography (EEG), and Ultrasound imaging have become more precise and informative due to sophisticated signal processing algorithms. These medical imaging modalities rely heavily on the ability to process, reconstruct, and interpret complex signals to provide clinical insights. Enhanced signal processing techniques have contributed to better diagnostic capabilities and advanced patient monitoring systems.

In audio and video processing, technologies like noise reduction, echo cancellation, and high-definition audio and video standards owe their capabilities to advancements in signal processing. The development of codecs like MP3, AAC, and H.264 have made digital media more accessible, allowing high-quality transmission and storage while maintaining manageable file sizes. These codecs employ sophisticated compression techniques that are made possible through advanced signal processing algorithms.

Industrial applications have also benefited. For instance, the development of smart sensors and automation systems in manufacturing relies on real-time signal processing to monitor and control complex processes accurately. Fault detection and predictive maintenance systems employ advanced signal processing techniques to analyze data from various sensors, thus preventing costly downtime and improving operational efficiency.

Continuing developments in quantum computing may represent another forthcoming revolution in signal processing. Quantum algorithms have the potential to solve certain types of problems significantly faster than classical algorithms, offering new possibilities for signal analysis and encryption.

Technological advancements have indubitably transformed signal processing into a more versatile and powerful discipline. By leveraging modern computing power, advanced algorithms, and new communication standards, the field has evolved to tackle increasingly complex problems across a variety of applications. As technology continues to advance, the potential for further innovations in signal processing remains vast, promising continued improvements across multiple domains.

2.4 The Impact of Legal and Ethical Considerations on Signal Processing

The domain of signal processing, integral to myriad applications ranging from telecommunications to medical diagnostics, is not immune to the pervasive influence of legal and ethical considerations. These factors shape the industry's landscape, governing the development, deployment, and sustenance of signal processing systems and technologies. The impact of these considerations is profound, influencing not only the legality of operations but also guiding moral imperatives to protect user rights, data privacy, and societal well-being.

Legal considerations in signal processing often revolve around data protection laws, intellectual property rights, regulatory compliance, and national security implications. The General Data Protection Regulation (GDPR) in the European Union, for example, imposes stringent requirements on how personal data can be collected, processed, and stored. Signal processing systems, particularly those involved in data-heavy applications like speech recognition and biometric identification, must adhere to these regulations to avoid severe penalties. Failure to comply with GDPR can result in fines amounting to 4% of a company's global annual turnover or €20 million, whichever is higher. This legal framework necessitates that systems are designed with privacy and data protection principles in mind, often leading to the incorporation of encryption methods and anonymization techniques to safeguard personal information.

Intellectual property rights also play a crucial role in shaping the signal processing landscape. Innovations in algorithms, methods, and hardware solutions must navigate through a maze of patents and trademarks to avoid infringements. Companies investing heavily in research and development seek to protect their inventions, which in turn propels the entire industry towards continual innovation and improvement. However, such protective measures can also stifle smaller entities and create monopolies, raising concerns about fair competition and access to essential technologies.

Ethical considerations extend beyond legal requirements, touching upon the broader implications of signal processing technologies on society. Privacy concerns are paramount, especially as signal processing applications increasingly pervade everyday life through smart devices, surveillance systems, and health monitoring tools. Ethical guidelines necessitate that these technologies do not infringe on individuals' privacy without consent. For example, facial recognition systems deployed in public spaces must balance the trade-off between security and individual rights, ensuring that they are not used for unwarranted surveillance or racial profiling.

Another critical ethical issue is the potential bias in signal processing algorithms. Machine learning and artificial intelligence, foundational to modern signal processing, are susceptible to biases present in training data. These biases can lead to discriminatory outcomes, such as disparate error rates for different demographic groups in voice recognition systems. Ethically responsible development mandates rigorous testing and validation against diverse datasets to minimize such biases and ensure equitable performance across all user categories.

Furthermore, the ethical implications of signal processing extend to its impact on employment and the workforce. Automated signal processing systems in industries like manufacturing and customer service can lead to job displacement, raising concerns about livelihoods and economic disparity. Ethical frameworks advocate for responsible innovation that includes considerations for job retraining and the development of new employment opportunities in sectors created by these technological advances.

National security is another pivotal aspect where legal and ethical considerations intersect with signal processing. Technologies capable of intercepting and analyzing communications, detecting signals from various sources, and ensuring cybersecurity are critical to national defense. However, the deployment and use of such technologies must abide by international laws and ethical guidelines to prevent misuse, abuse of power, and encroachment on civil liberties.

In conclusion, the intersection of legal and ethical considerations with signal processing underscores a complex yet necessary dimension of responsible technological innovation. Legal frameworks demand adherence to regulations that protect intellectual property and personal data, while ethical principles emphasize the need for fairness, privacy, and societal good. Navigating this multifaceted landscape requires a balanced approach, ensuring that advancements in signal processing reflect both legal compliance and ethical integrity, thereby fostering trust and promoting the universal benefits of technological progress.

3. Understanding Signal Processing Techniques

The advancing field of signal processing stands at the intersection of engineering, mathematics, and technology, forging pathways for innovations across numerous disciplines. It encompasses a wide range of techniques that manipulate signals—whether they be audio, video, sensor readings, or other data forms—to enhance, analyze, and efficiently transmit information. This chapter delves into the foundational and contemporary signal processing techniques that are pivotal in today's digital landscape. By exploring the theoretical underpinnings and practical applications, we aim to provide a comprehensive understanding that empowers further exploration and implementation of these critical methodologies. Through detailed examples and case studies, readers will gain insights into how signal processing not only solves complex problems but also drives technological advancements in communications, multimedia, biomedical engineering, and beyond.

3.1 The Basic Model: Input, Processing, Output

The foundational concept in computer science and information technology is often illustrated by the basic model of input, processing, and output. This model simplifies the complex interactions involved in computational processes and provides a framework for understanding how digital systems function. It conceptualizes the steps through which data is transformed into usable information, thereby facilitating a deeper comprehension of computing principles.

At the outset, the process begins with input. Input refers to the data or information that is fed into a computer system for processing. This data can originate from a variety of sources, such as user interactions through keyboards, mice, touchscreens, or external sensors, as well as automated feeds like network data packets or digitally stored information. The raw input data may be in numerous forms, including text, numbers, images, audio, or video. The crucial aspect of this step is the ability of the computer to accept and interpret this vast array of data types, preparing it for subsequent processing.

Once the input is gathered, the next phase is processing. During processing, the computer's central processing unit (CPU) or other processing units perform operations on the data as per the instructions provided by software programs or algorithms. This stage is the core of the computational process and involves the manipulation, transformation, and analysis of data to produce a meaningful output. Processing can encompass a wide range of activities, from simple arithmetic calculations and sorting to complex data analysis, statistical operations, image processing, and machine learning algorithms. The efficiency and speed of this stage are paramount, as they directly affect the performance and responsiveness of the system. During this phase, data may also be temporarily stored or retrieved from memory to enhance processing efficiency and manage larger datasets.

Following the completion of data processing, the final stage in this model is output. Output refers to the presentation of processed data in a comprehensible format. This can be visual, auditory, or tactile, depending upon the needs of the user and the nature of the processed data. Common forms of output include displayed text on a screen, printed documents, auditory signals from speakers, or even haptic feedback in a touch-enabled device. Additionally, output can connect to external systems, such as sending processed information over a network to another machine or storing it in databases for future retrieval and analysis. The quality and format of the output are vital because they determine how effectively the information can be used and interpreted by the end user.

These three stages, while distinct, are interdependent and often occur in an iterative or cyclical manner within a computing environment. Feedback loops are frequently employed to refine the accuracy and efficiency of each stage. For instance, output can be analyzed to detect errors or inconsistencies, leading to adjustments in the input phase or modifications in the processing algorithms.

Moreover, advancements in technology have continuously enhanced each aspect of this model. The development of more sophisticated input devices, faster and more capable processors, and higher-resolution output displays have greatly expanded the range and complexity of tasks that can be accomplished by computer systems. The integration of artificial intelligence and machine learning algorithms into the processing phase, for instance, has pushed the boundaries of what machines can understand and accomplish.

In conclusion, the basic model of input, processing, and output serves as a crucial abstraction that underlies the functioning of computer systems. It provides a structured and accessible way to understand how raw data is transformed into valuable information. This model is not merely theoretical but is practically implemented across a multitude of digital applications, from simple arithmetic tools to advanced supercomputing environments. Understanding this model facilitates a broader comprehension of computational systems and their vast potential in addressing diverse informational needs.

3.2 The Role of Feedback in Effective Signal Processing

Feedback is a pivotal component in the realm of signal processing, serving as a crucial mechanism for enhancing accuracy, efficiency, and overall system performance. In signal processing systems, feedback refers to the process by which a portion of the output signal is returned to the input, either directly or via some processing path. This mechanism is employed in a variety of applications including control systems, communication systems, and audio electronics among others. The role of feedback in effective signal processing encompasses numerous facets, from maintaining stability to improving the quality of the output signal.

To begin with, stability is one of the foremost considerations in any signal processing system. Feedback can significantly contribute to the stabilization of such systems. In negative feedback loops, the feedback signal opposes the input signal, thereby reducing the gain of the system and mitigating the risk of oscillations. This kind of feedback ensures that the system operates within its desired parameters, avoiding instability which could otherwise lead to unpredictable and often undesirable outcomes. A typical example can be found in amplifier circuits, where negative feedback is extensively used to stabilize gain and reduce distortion.

Another critical aspect wherein feedback plays a pivotal role is in the minimization of errors. In many signal processing systems, the presence of noise or other types of interference can lead to significant errors in the processed signal. Feedback mechanisms can be utilized to compare the output with a desired reference signal, and any detected discrepancies can be adjusted accordingly. This error correction capability is particularly valuable in communication systems where the integrity of transmitted data is paramount. By employing feedback, systems can dynamically adjust to minimize errors and enhance the fidelity of the received signal.

In adaptive signal processing, feedback is instrumental in the dynamic adjustment of system parameters. Adaptive systems are designed to operate in environments where signal characteristics can change over time. Feedback allows these systems to continually monitor their performance and make real-time adjustments to maintain optimal performance. For example, adaptive noise canceling headphones use feedback to constantly monitor ambient noise levels and adjust the noise cancelling algorithm, ensuring effective reduction of unwanted sounds over varying conditions.

Furthermore, feedback mechanisms are essential in the context of filter design, particularly in Infinite Impulse Response (IIR) filters. These filters employ feedback to achieve a desired frequency response, often achieving sharper cutoffs and more efficient filtering compared to their Finite Impulse Response (FIR) counterparts. The feedback in IIR filters enables them to use fewer coefficients to achieve the same performance level, which can be a crucial factor in applications where computational resources are limited.

In control systems, feedback loops are fundamental to achieving desired system behavior. The concept of Proportional-Integral-Derivative (PID) controllers is a prime example of how feedback is utilized to regulate system output. In such controllers, the feedback is used to compute the difference between a setpoint (desired value) and the current output, and adjustments are made based on proportional, integral, and derivative terms of this error. This ensures that the system quickly and accurately reaches and maintains the setpoint, despite disturbances or changes in system dynamics.

An important consideration in the design of feedback systems is the possibility of introducing instability if not properly managed. Positive feedback, where the feedback signal reinforces the input signal, can lead to exponential signal growth and potential system failure. Thus, careful design and analysis are required to ensure that feedback contributes positively to system performance without leading to instability.

In conclusion, feedback serves as a fundamental mechanism in effective signal processing, contributing to system stability, error minimization, adaptive performance, efficient filtering, and accurate control. By carefully designing feedback paths and understanding their impact on system behavior, engineers and scientists can harness the power of feedback to enhance the effectiveness and reliability of signal processing systems across a myriad of applications.

3.3 Overcoming Barriers to Signal Accuracy

In any signal processing system, achieving high accuracy is paramount for ensuring that the intended information is conveyed correctly from the sender to the receiver. Various barriers can impede signal accuracy, including noise, interference, distortion, and hardware limitations. Overcoming these barriers involves a combination of techniques and strategies that aim to maximize the integrity and fidelity of the transmitted signal.

Noise is one of the most prevalent factors affecting signal accuracy. It is any unwanted random addition to the signal that can distort the original information. Sources of noise can be both internal, such as thermal noise within the components of a system, and external, such as electromagnetic interference from other devices. One common method for mitigating noise is through the use of filtering techniques. Low-pass filters can remove high-frequency noise components, while high-pass filters can eliminate low-frequency noise. Additionally, advanced noise reduction algorithms, such as wavelet denoising and adaptive filtering, can dynamically adjust to varying noise levels to preserve the integrity of the signal.

Interference from other signals is another challenge. This can occur when multiple signals with similar frequencies overlap, leading to a phenomenon known as cross-talk. To combat this, techniques such as frequency hopping and spread spectrum can be employed. Frequency hopping involves rapid changes in the carrier frequency according to a sequence known to both the transmitter and receiver, thus minimizing the chance of interference. Spread spectrum techniques spread the signal over a wider frequency band, making it less susceptible to narrowband interference.

Distortion is another barrier impacting signal accuracy. It occurs when the signal's waveform changes shape as a result of the medium through which it is transmitted or the components used in the system. One way to address distortion is through the use of equalization techniques, which adjust the received signal to counteract the known effects of distortion before further processing or decoding. Linear equalizers operate by applying a filter that is the inverse of the distortion, while adaptive equalizers dynamically adjust their characteristics based on the changing signal environment.

Hardware limitations can also play a significant role in reducing signal accuracy. Imperfections in components, such as amplifiers and analog-to-digital converters, can introduce errors. To mitigate these issues, careful design and selection of high-quality components are essential. Moreover, calibration procedures can help correct any deviations in component performance over time. In highly sensitive applications, redundancy and error correction methods, such as the use of error-correcting codes, can be introduced. These codes add extra bits to the transmitted data that the receiver can use to detect and correct any errors that may have occurred during transmission.

Digital signal processing (DSP) offers several techniques to enhance signal accuracy by converting the signal from analog to digital form, where sophisticated algorithms can be applied. Techniques such as Fast Fourier Transform (FFT) allow for the efficient analysis and filtering of signals. DSP can also implement error detection and correction algorithms more effectively than analog systems. Additionally, digital systems often have higher immunity to noise and distortion compared to their analog counterparts.

Error detection and correction mechanisms are vital for maintaining signal accuracy in digital communications. Parity checks, checksums, and cyclic redundancy checks (CRC) are simple yet effective means of detecting errors. More advanced error correction techniques, such as Hamming codes, Reed-Solomon codes, and convolutional codes, not only detect but also correct errors without requiring retransmission of the data. Forward error correction (FEC) is particularly beneficial in environments with high error rates, such as satellite communications, where retransmissions are costly or impractical.

Despite these techniques, environmental factors can still pose challenges to signal accuracy. Signal degradation due to attenuation, multipath propagation, and Doppler shift must be addressed, particularly in wireless communication systems. Techniques like adaptive modulation, where the modulation scheme is adjusted according to the channel conditions, and multiple-input multiple-output (MIMO) systems, which use multiple antennas to leverage spatial diversity, can significantly enhance signal robustness and accuracy.

In conclusion, overcoming barriers to signal accuracy in complex systems requires a multifaceted approach. Through the application of filtering techniques, interference management, distortion correction, careful hardware design, digital signal processing, and error correction methods, it is possible to significantly enhance the accuracy and reliability of signals. Each barrier presents its own unique challenges, but with the appropriate strategies and technologies, these obstacles can be effectively mitigated, ensuring that accurate and reliable communication is maintained.

3.4 Adapting Signal Processing Strategies to Different Applications

Signal processing, a field dealing with the analysis, manipulation, and synthesis of signals, finds applications ranging from audio and speech processing to telecommunications, biomedical engineering, radar systems, and beyond. The strategies employed in signal processing must adapt to the specific requirements, constraints, and goals of these diverse applications. This adaptation involves a meticulous selection of algorithms, consideration of hardware limitations, and a keen understanding of the signal characteristics pertinent to each application.

One key aspect of adapting signal processing strategies is the nature of the signal itself—whether it's analog or digital. For instance, in audio signal processing, signals are typically acquired in analog form, such as sound waves. They are then converted to digital form using analog-to-digital converters (ADCs) to facilitate more sophisticated manipulation. The choice of ADC, including its sampling rate and resolution, must match the application's requirements to faithfully capture the signal's nuances. Failure to adapt the sampling rate appropriately, guided by the Nyquist-Shannon sampling theorem, could lead to aliasing, compromising the signal's integrity.

Data compression is another area where signal processing strategies must be tailored. In telecommunications, bandwidth is often limited, necessitating the use of compression algorithms such as those used in codecs like MP3 for audio or H.264 for video. Signal fidelity must be balanced with compression efficiency to ensure quality transmission within the available bandwidth. Contrastingly, in medical imaging like Magnetic Resonance Imaging (MRI), the paramount concern is often maintaining the highest resolution and fidelity possible to ensure accurate diagnostics, necessitating different compression techniques or sometimes foregoing lossy compression altogether.

The processing of signals in noisy environments also demands unique considerations. For example, in radar systems, signals are frequently obscured by substantial noise and interference. Adaptive filtering techniques such as the Kalman Filter or Wiener Filter are employed to differentiate the desired signal from noise, enhancing detection and accuracy. Similarly, in underwater acoustics, where sonar is deployed, signal degradation and multipath propagation create challenges requiring robust, adaptive signal processing strategies to extract useful information from noise-laden returns.

Real-time processing requirements further influence the choice of signal processing strategies. Applications such as real-time audio effects processing or live video streaming impose stringent latency constraints. Here, computational efficiency becomes crucial. Fast Fourier Transform (FFT) algorithms, which significantly reduce the computational burden of Fourier Transforms, are often leveraged in these scenarios to enable real-time processing without violating latency constraints.

Adapting signal processing strategies also entails addressing hardware constraints. In embedded systems, such as hearing aids or sensors in Internet of Things (IoT) devices, processing power and memory are limited. Lightweight algorithms or hardware accelerators, like Digital Signal Processors (DSPs) or Field-Programmable Gate Arrays (FPGAs), may be employed to meet these limitations while still achieving the necessary signal processing tasks.

Multidimensional signal processing, relevant in applications like image and video processing, requires strategies that can handle the complexity of managing data across multiple dimensions. For example, image processing algorithms such as convolutional neural networks (CNNs) are tailored to recognize patterns and features in two-dimensional data. The adaptation here involves selecting appropriate network architectures and training methodologies that account for the spatial characteristics of images.

In biomedical signal processing, considerations such as patient safety and data privacy introduce additional layers of complexity. Electrocardiograms (ECGs), for instance, have unique characteristics such as distinct wave patterns (P, QRS, and T waves) requiring specialized detection and classification algorithms. Adaptive algorithms that can monitor and process these signals in real-time, detect anomalies, and alert healthcare providers are crucial. Furthermore, ensuring that these algorithms operate on secure platforms to protect sensitive medical data is an indispensable aspect of adapting signal processing strategies in this domain.

In conclusion, adapting signal processing strategies to different applications is a multi-faceted endeavor. It requires a deep understanding of the signal characteristics, the specific application requirements, and the constraints imposed by the operating environment. From selecting appropriate algorithms and hardware to accounting for real-time processing needs and ensuring robustness in noisy conditions, the adaptation process is vital for the successful application of signal processing techniques across diverse domains.

4. Signal Processing and Technological Advancement

The evolution of technology has always been tightly intertwined with advancements in signal processing. This chapter delves into how signal processing has shaped the landscape of modern technology, influencing everything from telecommunications to medical diagnostics. As digital transformation accelerates, the importance of efficient and sophisticated signal processing techniques cannot be overstated. We will explore fundamental principles, groundbreaking innovations, and the practical applications that drive today's technological progress. By examining both historical milestones and current trends, we aim to offer a comprehensive understanding of how signal processing is pivotal in fostering continued innovation and addressing emerging challenges in various fields.

4.1 The Role of Signal Processing in Driving Technological Innovation

Signal processing is a critical field that serves as the backbone for numerous technological advancements. This domain encompasses the analysis, manipulation, and interpretation of signals, which can be any form of data that conveys information. Signals are ubiquitous; they include everything from audio and visual data to sensor readings and communication signals. Through sophisticated algorithms and computational techniques, signal processing enables the extraction of valuable information from these raw signals, consequently driving technological innovation across various sectors.

One of the primary areas where signal processing has spurred innovation is in telecommunications. The modern communication systems we rely on—from mobile phones to the internet—are built upon complex signal processing techniques. These techniques include modulation and demodulation, error detection and correction, and data compression. For example, the transformation of an analog voice signal into a digital format, its transmission over long distances, and its subsequent reconstruction into an audible form—all these steps involve extensive signal processing. This transformation is essential for high-quality voice and video calls, seamless streaming services, and reliable data transfer, forming the core of how we communicate today.

In addition to telecommunications, signal processing has had a transformative impact on the field of medical imaging. Technologies such as Magnetic Resonance Imaging (MRI), Computed Tomography (CT), and ultrasound rely heavily on signal processing to interpret and visualize internal body structures. These imaging techniques involve the acquisition of raw data through sensors, which is then systematically processed to generate detailed images. Advanced algorithms enhance image clarity, reduce noise, and even enable 3D reconstructions, allowing for more accurate diagnoses and better patient outcomes. Signal processing also plays a pivotal role in the development of wearable health devices, which monitor vital signs and detect anomalies in real-time, thus contributing to personalized medicine and proactive healthcare.

Signal processing is also instrumental in the advancement of machine learning and artificial intelligence (AI). Machine learning algorithms often require preprocessing of data to filter noise, normalize values, and extract pertinent features—a process fundamentally rooted in signal processing principles. Signal processing techniques such as Fourier transforms, wavelets, and spectral analysis are frequently employed to prepare data for machine learning applications. This preprocessing is crucial for the performance and accuracy of AI models, impacting areas like speech recognition, natural language processing, and image classification. Consequently, signal processing underpins many of the intelligent technologies that are becoming integral to our daily lives, from virtual assistants to autonomous vehicles.

Another domain where signal processing plays a vital role is in environmental monitoring and remote sensing. Modern satellite systems equipped with advanced sensors collect vast amounts of data about Earth's atmosphere, oceans, and land surfaces. Signal processing algorithms are used to interpret these data, enabling accurate weather forecasting, climate change monitoring, and disaster management. For instance, the detection of changes in sea surface temperatures or the analysis of deforestation patterns involves complex signal processing procedures. The insights derived from these analyses support informed decision-making and policy formulation, aiding in the preservation and sustainable management of natural resources.

Moreover, the field of audio and video technology has seen revolutionary changes due to advancements in signal processing. Audio signal processing techniques facilitate noise reduction, echo cancellation, and sound enhancement, which are crucial for high-fidelity audio systems, hearing aids, and teleconferencing equipment. Similarly, video signal processing has enabled the development of high-definition and 4K video technologies, as well as innovations in video compression standards like H.264 and H.265. These advancements have significantly improved the quality of multimedia content delivery, enriching user experiences in entertainment, education, and professional environments.

Finally, signal processing has enabled significant breakthroughs in the realm of cybersecurity. Techniques such as cryptographic signal processing are employed to ensure the confidentiality, integrity, and authenticity of data transmitted over networks. These techniques protect sensitive information from being intercepted or tampered with by malicious entities. Additionally, signal processing methods are used in anomaly detection systems to identify unusual patterns that may indicate security breaches. This vigilance is crucial for protecting critical infrastructures, financial systems, and personal data from cyber threats.

In summary, signal processing is a foundational element that drives technological innovation across a multitude of fields. Its applications in telecommunications, medical imaging, machine learning, environmental monitoring, audio and video technology, and cybersecurity underscore its pervasive influence. By enabling the efficient and effective extraction of information from raw data, signal processing continues to propel advancements that shape the technological landscape and enhance our everyday lives.

4.2 Measuring the Impact of Signal Processing on Performance

Measuring the impact of signal processing on performance is an essential aspect of evaluating the effectiveness and efficiency of a communication system. Signal processing encompasses a range of techniques used to manipulate, transform, and analyze signals to improve their quality or to extract useful information. Understanding how these signal processing methods influence performance requires a comprehensive assessment of various factors such as signal quality, computational complexity, power consumption, and overall system reliability.

Firstly, the quality of processed signals is often gauged using metrics like Signal-to-Noise Ratio (SNR), Bit Error Rate (BER), and Mean Squared Error (MSE). SNR measures the level of the desired signal relative to the background noise. An effective signal processing technique will result in a higher SNR, indicating that the signal is clearer and more distinguishable from noise. BER, on the other hand, quantifies the number of bit errors in a transmitted data stream. Lower BER values reflect a more reliable transmission, pointing to effective error correction and filtering mechanisms employed by the signal processing algorithms. MSE gives an idea of the average squared difference between original and processed signals, providing insights into the precision and fidelity of the processing technique.

Computational complexity is another critical factor determining the practicality of a signal processing method. Techniques that are computationally intensive may yield high fidelity but can be impractical in real-time or resource-constrained applications. Evaluating computational complexity involves assessing the algorithm's time complexity and the number of required operations. Techniques like Fast Fourier Transform (FFT) are highly valued in signal processing for their balance between performance enhancement and manageable computational complexity. An optimal signal processing algorithm achieves the desired signal quality improvements without imposing an excessive computational burden, making it feasible for implementation in various real-time systems.

Power consumption further influences the overall assessment of signal processing techniques. In battery-operated and portable devices, algorithms that deliver high-quality signal enhancement at lower power usage are highly preferred. Measuring the energy efficiency of a signal processing technique involves monitoring the power consumption during its operation and finding the trade-offs between energy use and performance gains. Advanced techniques like adaptive filtering, which dynamically adjusts its parameters based on the signal conditions, can offer significant power savings while maintaining desired performance levels.

System reliability and robustness add another layer of complexity to the evaluation process. The impact of signal processing on performance must account for how well the system can handle adverse conditions, such as signal interference, multipath propagation, or hardware failures. Robust signal processing algorithms are designed to maintain performance levels even under challenging conditions. Techniques such as redundancy, error correction codes, and adaptive algorithms enhance reliability and enable systems to recover from or compensate for various types of signal degradation. Measuring performance in this context involves simulating or experimentally subjecting the system to challenging scenarios and observing how well the signal processing methods cope with them.

Lastly, the practical implementation and scalability of signal processing techniques play a crucial role. Techniques that are theoretically sound must also be capable of practical deployment in real-world systems. This involves considering the hardware constraints, potential implementation costs, and the ability to scale the technique across different platforms and applications. Performance metrics in this domain include throughput, latency, and the system's ability to handle increased data rates or more complex signal environments.

In summary, evaluating the impact of signal processing on performance is a multi-faceted endeavor that spans signal quality, computational complexity, power consumption, system reliability, and practical implementation. Each of these factors contributes to a holistic understanding of how well signal processing techniques perform in enhancing communication systems. An optimal approach balances these elements to achieve the desired improvements without compromising on practicality and efficiency, paving the way for more robust and adaptive communication technologies.

4.3 Signal Processing as a Competitive Advantage

Signal processing is an interdisciplinary domain that intersects with electrical engineering, applied mathematics, and computer science, among other fields. Its fundamentals involve the analysis, synthesis, and modification of signals, which are essentially any time-varying or spatial-varying quantities. These signals might represent sounds, images, biological data, or any other form of information that can be converted into digital or analog formats. The deep understanding and effective manipulation of these signals can grant a substantial competitive edge across various industries.

The core advantage of advanced signal processing lies in its ability to enhance efficiency, precision, and performance in multifaceted ways. For example, in telecommunications, sophisticated signal processing techniques are employed to improve the clarity and reliability of data transmission. Systems utilizing advanced modulation and error correction techniques can deliver higher bandwidth and lower error rates, even over long distances and noisy channels. This not only translates to a better user experience but also lesser operational costs due to reduced need for retransmission and lower infrastructure expenditures.

In the realm of consumer electronics, signal processing has revolutionized everything from audio and visual experiences to smart devices' functionalities. Enhanced algorithms convert raw audio signals into high-fidelity sound in speakers and headphones, improve image and video quality in cameras and displays, and enable voice-recognition capabilities in smartphones and home assistants. Companies that can integrate signal processing more effectively into their products can thus offer richer, more intuitive, and higher-quality user experiences, setting themselves apart from competitors.

Medical technology is another area where signal processing has a profound impact. In diagnostics, imaging techniques such as MRI, CT scans, and ultrasound heavily rely on signal processing to convert raw data into coherent and interpretable images. Advanced algorithms can enhance the resolution, reduce noise, and even identify subtle anomalies that might go unnoticed by human experts. In patient monitoring, real-time signal processing of physiological data such as ECG, EEG, and blood pressure can provide immediate insights, allowing for timely interventions. Innovations in this sphere can mean the difference between early detection of critical conditions and missed diagnostic opportunities, positioning firms with superior signal processing capabilities at the forefront of medical advancement.

Further extending its utility, signal processing is crucial in the automotive industry, particularly within the realm of autonomous driving. The myriad sensors equipped in modern vehicles, including LiDAR, radar, and cameras, generate enormous amounts of data. Robust signal processing algorithms are essential to digesting this data, making sense of the surroundings, and ensuring safe and efficient navigation. Companies that excel in this area not only enhance the safety and performance of their autonomous systems but also expedite the journey towards fully autonomous vehicles, thus capturing market leadership.

Moreover, signal processing technologies possess immense importance in security and surveillance. Facial recognition, fingerprint analysis, and behavioral biometrics have all been revolutionized through advanced signal processing. In military and defense applications, signal processing aids in radar and sonar systems, electronic warfare, and secure communications, providing tactical superiority through enhanced situational awareness and secure, reliable data exchange.

Finally, the financial sector has become increasingly dependent on signal processing techniques for high-frequency trading, risk management, and market analysis. By converting streams of market data into actionable signals through complex mathematical models and real-time analytics, traders and analysts can uncover patterns, anticipate market movements, and execute trades with microsecond precision. This not only maximizes profitability but also mitigates risk, offering a decisive edge in the highly competitive world of finance.

In conclusion, signal processing as a competitive advantage is characterized by its ability to transform raw data into actionable insights, improve system efficiencies, and enhance overall product and service quality across diverse industries. Its ever-evolving methodologies contribute significantly to innovation, enabling entities that harness its power effectively to lead in their respective markets. Whether it's enhancing consumer experiences, advancing medical diagnostics, or pioneering autonomous technologies, the strategic application of signal processing holds the potential to redefine industry standards and deliver unsurpassed competitive advantage.

4.4 Case Studies of Innovation-Driven Signal Processing Strategies

In recent years, the field of signal processing has witnessed unprecedented advancements due to innovative strategies that leverage cutting-edge technology and sophisticated algorithms. These advancements have been propelled by the need to tackle rapidly evolving challenges in various industries. This exploration delves into four case studies where innovation-driven signal processing strategies have made significant impacts.

Case Study 1: Medical Imaging Enhancemen

Medical imaging is pivotal in diagnosing and treating ailments. Traditional signal processing techniques have limitations, particularly in producing clear and precise images. Recent innovations, however, have revolutionized this field. One key development is the use of deep learning algorithms to enhance the quality of medical images, such as MRI and CT scans.

Researchers have developed convolutional neural networks (CNNs) specifically tailored for image enhancement. These networks learn to recognize patterns and structures in medical images, effectively reducing noise and improving resolution. For example, in the detection of tumors, a CNN can outperform traditional methods by providing clearer images, which allow radiologists to make more accurate diagnoses. This convergence of machine learning and signal processing exemplifies how innovative strategies are transforming healthcare.

Case Study 2: Adaptive Noise Cancellation in Telecommunications

Telecommunications systems must often operate in environments with significant background noise, which can degrade the quality of transmitted signals. Traditional noise cancellation techniques, such as simple filtering, fall short in dynamically changing environments. Adaptive noise cancellation, however, has emerged as a robust solution.

Adaptive algorithms, such as the Least Mean Squares (LMS) algorithm, dynamically adjust filters in real-time to minimize noise. These algorithms use feedback loops to continuously learn and adapt to the noise environment. This technology is particularly beneficial in mobile communications and voice-controlled devices, where maintaining clear audio quality is paramount. For instance, adaptive noise cancellation in smartphones enhances call quality by reducing wind and traffic noise, leading to a more pleasant user experience.

Case Study 3: Smart Grid Energy Management

With the increasing adoption of renewable energy sources, managing energy in smart grids has become a complex challenge. Signal processing strategies have been integral in optimizing the distribution and consumption of energy. One innovative approach is the application of predictive analytics to forecast energy demand and supply.

By employing machine learning models, energy consumption patterns can be predicted based on historical data. These models process large volumes of data from various sensors and smart meters to predict future energy needs accurately. Consequently, utilities can adjust their operations to ensure efficient energy distribution, reduce wastage, and integrate renewable sources seamlessly. For instance, predictive analytics enables energy providers to balance demand and supply more effectively during peak hours by engaging in demand-response programs.

Case Study 4: Autonomous Vehicle Navigation

Autonomous vehicles rely heavily on signal processing to interpret and react to their environment. Traditional approaches hinge on pre-defined algorithms with limited adaptability. However, innovative strategies have incorporated advanced signal processing techniques driven by artificial intelligence (AI) to enhance navigation and safety.

One breakthrough is the use of sensor fusion, where data from various sensors such as LiDAR, cameras, and radar are combined to create a comprehensive understanding of the vehicle's surroundings. Machine learning algorithms process this fused data in real-time, enabling the vehicle to make informed decisions. Deep reinforcement learning, in particular, allows autonomous systems to learn from interactions with the environment, improving their navigation capabilities over time. A practical example is Tesla's Autopilot system, which uses an array of sensors and AI-driven signal processing to achieve semi-autonomous driving.

Conclusion

These case studies underscore the transformative influence of innovation-driven signal processing strategies across diverse domains. From enhancing medical imaging to enabling autonomous vehicles, these advancements are not only improving existing systems but also paving the way for new applications and industries. The key takeaway is the symbiotic relationship between innovation and signal processing; as technology evolves, so does our ability to tackle increasingly complex problems through more sophisticated methods. As we continue to push the boundaries, the future promises even more groundbreaking developments in this dynamic field.

5. Strategic Planning for Signal Processing

Strategic planning is an essential discipline that bridges the gap between high-level objectives and the practical implementation of signal processing systems. In the rapidly evolving landscape of technology, the significance of a well-articulated strategy cannot be overstated. Precise foresight and comprehensive planning are fundamental to maintaining competitiveness and achieving long-term goals.

This chapter delves into the intricacies of strategic planning within the context of signal processing, an area characterized by its dynamic algorithms, complex data structures, and crucial applications. We will explore how to align various methodologies and tools with overarching organizational goals, ensuring that signal processing initiatives are not only technically sound but also mission-driven and commercially viable.

The discussion will cover key aspects such as defining clear objectives, assessing current capabilities, forecasting technological trends, and devising a flexible yet robust plan of action. Emphasis will be placed on practical frameworks and models that facilitate decision-making and prioritize the allocation of resources efficiently.

Through a combination of theoretical insights and practical examples, this chapter aims to equip readers with the knowledge and skills necessary to navigate the multifaceted challenges of strategic planning in the realm of signal processing. Whether you are a seasoned engineer, a project manager, or a decision-maker, the strategies presented will help you foster innovation, enhance performance, and ultimately, achieve sustained success in your signal processing endeavors.

5.1 Developing a Comprehensive Signal Processing Strategy

Creating an effective signal processing strategy is essential for the successful extraction, manipulation, and analysis of information from various signal sources. It requires a comprehensive approach that takes into consideration multiple factors including the nature of the signal, the objectives of the processing, and the constraints of the system. Below is an elucidation of the elements involved in developing a comprehensive signal processing strategy.

The first step in formulating a robust signal processing strategy involves a thorough understanding of the signal characteristics and the application requirements. This often necessitates a detailed analysis of the signal's origin, its statistical properties, and its frequency content. Signals can be analog or digital, continuous or discrete, and each type demands specific processing techniques. For digital signals, considerations such as sampling rate, quantization, and modulation must be addressed to ensure the integrity of the captured information.

Once the signal's nature is comprehensively understood, the next stage is defining the objectives of the processing. These can range from basic tasks like filtering and amplification to more complex functions like feature extraction, pattern recognition, and predictive analysis. The goals could be single-faceted or multifaceted: some applications may simply require noise reduction, while others might need detailed spectral analysis to identify specific signal components or trends over time.

A pivotal aspect of developing a strategy is selecting appropriate algorithms and methodologies that meet the defined objectives. Signal processing encompasses a wide array of techniques including time-domain analysis, frequency-domain analysis, and time-frequency domain analysis. Each methodology has advantages and specific applications. For instance, Fourier Transform is widely utilized for converting a time-domain signal into its frequency components, which is beneficial in harmonic analysis and filtering. Wavelet Transform, on the other hand, offers advantages in localizing both time and frequency components simultaneously and is particularly useful in transient signal analysis.

Moreover, the implementation of signal processing requires a critical assessment of the hardware and software capabilities. The choice between using general purpose processors, Digital Signal Processors (DSP), or Field-Programmable Gate Arrays (FPGA) is dictated by the real-time processing requirements, power constraints, and system scalability. Software environments and toolboxes such as MATLAB, Simulink, and Python libraries (e.g., SciPy and NumPy) provide robust platforms for developing and testing processing algorithms before deploying them into practical hardware setups.

In the subsequent phase, validation and verification procedures are instituted to ensure that the strategies perform as anticipated under varied conditions. This involves rigorous testing using simulated data and real-world signals to verify the robustness and accuracy of the algorithms. Techniques such as cross-validation, error analysis, and sensitivity analysis are vital in this stage to identify any potential weaknesses or limitations in the strategy.

Furthermore, the strategy should encompass adaptive and learning mechanisms to enhance the signal processing framework. With advancements in machine learning and artificial intelligence, adaptive filtering and neural networks can be incorporated to offer better performance in dynamic and non-stationary signal environments. Such adaptive systems can automatically adjust their parameters based on the incoming signal characteristics, leading to improved accuracy and efficiency over time.

Lastly, the integration of the signal processing strategy into the overall system architecture must be planned meticulously. This ensures that the signal processing components interface seamlessly with other system modules such as data acquisition, storage, and user interface. Proper documentation and modular design principles aid in the maintainability and scalability of the system, allowing for future upgrades and modifications with minimal disruptions.

In conclusion, developing a comprehensive signal processing strategy is a multifaceted task that involves understanding the signal intricacies, defining clear objectives, selecting suitable algorithms, validating performance, incorporating adaptive mechanisms, and ensuring seamless integration into the broader system. Approaching this process systematically not only enhances the efficiency and accuracy of signal processing but also ensures that it meets the evolving demands of contemporary applications.

5.2 Aligning Processing Goals with Technological Objectives

Aligning processing goals with technological objectives is a critical aspect of efficient and effective business operations. Achieving this alignment ensures that the technological capabilities within an organization are leveraged to meet and exceed the desired outcomes of processes. This synergy between process goals and technological objectives drives performance improvements, cost reductions, innovation, and competitive advantage.

The first step in aligning processing goals with technological objectives is to thoroughly understand the specific business goals and process requirements. This understanding involves identifying the key outcomes that the organization aims to achieve through its processes. These goals could range from improved customer satisfaction, faster turnaround times, enhanced product quality, increased operational efficiency, to reduced costs. Organizations must conduct a detailed analysis of their existing processes to pinpoint areas that require improvement and determine the functional requirements needed to support these processes.

Once processing goals are clearly defined, the next step is to map these goals to appropriate technological solutions. This involves a comprehensive evaluation of the organization's technological infrastructure and capabilities. Modern technologies, such as automation, cloud computing, data analytics, artificial intelligence, and Internet of Things (IoT), offer transformative potential in enhancing process efficiencies. Selecting the right technology depends on the specific needs of the process and how well the technology can meet those needs. For instance, if a key goal is to improve data accuracy and speed in decision-making processes, implementing advanced data analytics tools and AI-driven decision support systems might be most effective.

Incorporating technology into process workflows requires meticulous planning and execution. One of the vital aspects includes ensuring that the chosen technology is scalable and adaptable. Business environments are dynamic, and the chosen technological solutions must be flexible enough to evolve with changing business needs. Additionally, integration with existing systems is crucial to avoid disruptions and ensure seamless operations.

Integration allows for the smooth flow of information across various departments, thereby enhancing overall process efficiency.

To effectively align processing goals with technological objectives, it is also essential to foster a culture of continuous improvement and innovation. This involves encouraging employees across all levels to embrace new technologies and adopt best practices in process management. Providing adequate training and development programs ensures that the workforce is proficient in using the new technologies and fully understands the benefits they bring to the process goals. Leadership plays a crucial role here, as it must champion the integration of technology, drive the vision forward, and address any resistance to change.

Furthermore, performance metrics and evaluation frameworks are critical to measure the success of the alignment efforts. Establishing key performance indicators (KPIs) that directly link process performance with technological capabilities helps in monitoring progress and making informed decisions. Metrics such as turnaround times, error rates, productivity levels, customer satisfaction scores, and cost savings provide tangible evidence of how well the technological objectives are meeting the processing goals. Regular reviews and adjustments based on these metrics ensure that the alignment remains effective and relevant over time.

Lastly, a clear communication strategy must be in place to articulate the vision, goals, and benefits of aligning processing goals with technological objectives. Transparent communication ensures that all stakeholders are on the same page and fosters a collaborative environment. Engaging all stakeholders, including employees, management, customers, and suppliers, in the planning and implementation stages can provide valuable insights and enhance the overall success of the alignment efforts.

In conclusion, aligning processing goals with technological objectives is a multifaceted approach that requires careful planning, execution, and continuous evaluation. By understanding business goals, selecting appropriate technologies, fostering a culture of innovation, and establishing robust performance metrics, organizations can achieve significant improvements in process efficiency and overall business performance. The alignment not only drives operational excellence but also positions the organization to adapt and thrive in an ever-evolving technological landscape.

5.3 Strategic Implementation of Signal Processing Initiatives

Signal processing initiatives are critical in various fields, including telecommunications, aerospace, and medical instrumentation, where they improve the interpretation and manipulation of data signals. Effective strategic implementation of these initiatives involves several structured phases to ensure they deliver the intended benefits and remain adaptable to future developments.

The first phase is the **Conceptualization and Planning** stage. Here, strategic goals and objectives are defined based on the specific needs and conditions of the organization or project. This stage requires a deep understanding of both the current capabilities and the potential improvements signal processing can bring. Stakeholders, including technical experts, project managers, and end-users, collaborate to set clear, measurable objectives, such as enhancing signal clarity, reducing noise, or improving data compression techniques. A feasibility study critically examines resource availability, technological constraints, and market demands to establish a realistic scope and timeline for the initiative.

Next comes the **Design and Development** phase. During this phase, theoretical models and algorithms are created based on the requirements outlined in the planning stage. Engineers and scientists work on developing algorithms that can efficiently process signals while meeting performance criteria such as speed, accuracy, and energy consumption. This stage often involves prototyping and iterative testing, where data from real-world scenarios is used to refine and optimize the algorithms. Techniques such as machine learning and artificial intelligence can be employed to enhance signal processing abilities, particularly in predicting and adapting to changing conditions.

The **Implementation stage turns** designs into practical solutions. Hardware and software infrastructure that will run the signal processing algorithms are developed or upgraded. For instance, Field-Programmable Gate Arrays (FPGAs) or Digital Signal Processors (DSPs) might be utilized for high-speed, real-time processing requirements. The implementation phase also includes rigorous testing in controlled environments to validate the performance of the systems under various conditions, ensuring they meet all specified technical and functional standards.

Upon successful testing, the initiative moves to the **Deployment** stage, where the signal processing systems are integrated into the operational environment. This phase involves collaboration with other departments, such as IT, operations, and customer support, to ensure seamless integration. Training programs may be conducted to familiarize users and maintainers with the new systems. Continuous monitoring and logging are set up to collect data on system performance and user feedback, helping to identify any potential issues early.

The final phase is **Evaluation and Maintenance**. This stage involves the continuous assessment of the signal processing systems. Key performance indicators (KPIs) and metrics defined in the planning stage are reviewed to determine the success of the initiative. Data collected during the deployment phase are analyzed to pinpoint strengths and weaknesses. Regular maintenance and updates are critical to ensure long-term viability, as signal processing technology rapidly evolves. This phase also includes scalability assessments to determine how the system can adapt to increased loads or new functionalities.

In sum, the strategic implementation of signal processing initiatives is a multifaceted process requiring careful planning, in-depth development, rigorous testing, seamless deployment, and ongoing evaluation. Each stage involves its own set of challenges and requires a collaborative effort across multiple disciplines to ensure the initiative is successful and sustainable. Through meticulous execution, signal processing initiatives significantly enhance the capability to interpret and manipulate data signals, driving improvements in various critical applications.

5.4 Adapting Signal Processing Strategies During Technological Changes

The landscape of signal processing has undergone significant transformations due to rapid technological advancements. Adapting signal processing strategies effectively amid these changes is crucial for both maintaining and enhancing system performance. This intricate process involves updating algorithms, integrating new hardware, and ensuring compatibility with emerging technologies. Each of these aspects requires detailed attention to maintain coherence, accuracy, and efficiency.

Firstly, updating algorithms is a fundamental aspect of adapting signal processing strategies. Algorithms designed for signal processing must evolve concurrently with technological advancements to leverage new capabilities and meet modern requirements. For example, traditional Fast Fourier Transform (FFT) algorithms have been refined to improve computational efficiency and accuracy in various applications, from telecommunications to healthcare. As new techniques in machine learning and artificial intelligence emerge, integrating these into signal processing can offer significant advantages. Machine learning algorithms, for instance, can improve signal detection, classification, and prediction by learning from data patterns and making informed decisions. The integration of such algorithms can also help in noise reduction, echo cancellation, and other complex signal processing tasks, thus enhancing overall system performance.

Secondly, the progression in hardware technology plays a pivotal role in shaping signal processing strategies. The evolution from analog to digital systems marked a significant shift, enabling more precise and flexible signal manipulation. Today, advancements in microprocessor technology, Field-Programmable Gate Arrays (FPGAs), and Graphic Processing Units (GPUs) provide immense computational power, which supports complex signal processing tasks in real-time. Utilizing these hardware advancements necessitates adapting signal processing algorithms to fully exploit the capabilities of the new hardware. For example, FPGAs can be tailored for specific signal processing tasks, making them highly efficient for tasks such as high-speed filtering and real-time data acquisition. Similarly, GPUs, with their parallel processing capabilities, are becoming crucial for handling the intensive computational demands of modern signal processing algorithms, particularly in applications involving large datasets and machine learning.

Compatibility with emerging technologies is another critical aspect to consider. Modern signal processing applications often need to interface with other technologies, such as Internet of Things (IoT) devices, 5G networks, and advanced sensor systems. Ensuring that signal processing strategies can seamlessly integrate with these technologies requires an understanding of their specific requirements and constraints. For example, IoT devices often operate on low power and have limited computational resources, thus demanding highly efficient signal processing algorithms that can function within these constraints. Similarly, 5G networks promise high data rates and low latency, which necessitate advanced signal processing techniques to manage the increased data flow and maintain communication quality. Adaptation in this context might involve developing new protocols or optimizing existing ones to ensure robust and efficient signal processing.

One notable trend in the adaptation process is the shift towards software-defined architectures. This approach allows for greater flexibility and scalability in signal processing systems by decoupling hardware from software. With software-defined radio (SDR) systems, for instance, signal processing tasks traditionally handled by hardware components are executed in software. This flexibility enables quicker updates and modifications to signal processing strategies to accommodate technological changes without the need for extensive hardware overhauls. Additionally, software-defined architectures facilitate the deployment of standardized platforms, which can lower costs and simplify maintenance.

Moreover, open-source software tools and libraries have become invaluable in the adaptation of signal processing strategies. Tools such as GNU Radio, a free software development toolkit, provide a flexible environment for developing and implementing signal processing algorithms. These tools not only accelerate the adaptation process but also encourage collaboration and innovation, as researchers and engineers can share and improve upon each other's work.

In conclusion, adapting signal processing strategies during technological changes involves a comprehensive approach that includes updating algorithms, integrating new hardware, and ensuring compatibility with emerging technologies. By addressing these aspects, it is possible to maintain and enhance system performance, thereby meeting the evolving demands of modern applications. Continuous learning and adaptation are key to thriving in this dynamic field, ensuring that signal processing systems remain robust, efficient, and relevant in the face of ongoing technological advancements.

6. Digital Signal Processing Techniques

In the ever-evolving landscape of modern technology, digital signal processing (DSP) stands as a cornerstone, shaping the way we interact with and interpret data. This chapter delves into the intricate yet fascinating world of DSP, exploring the algorithms, methodologies, and innovations that enable the transformation and enhancement of digital signals. As the demand for high-quality and efficient digital communication continues to surge, understanding DSP techniques becomes not just beneficial, but essential.

The journey begins with fundamental principles that ground our understanding of how digital signals are represented and manipulated. From the mathematical foundations to the practical applications, we will uncover the multi-faceted aspects of DSP, emphasizing how these techniques are employed to address real-world challenges. Whether it be in telecommunications, audio processing, medical imaging, or any number of other fields, the advancements in DSP contribute significantly to the developments that characterize our digital age.

As you venture through this chapter, you will encounter a blend of theoretical insights and practical examples designed to illustrate the power and versatility of digital signal processing. Prepare to engage with a topic that not only underpins many of today's technologies but also promises to be a driving force in tomorrow's innovations.

6.1 Implementing Effective Digital Filters and Algorithms

Implementing effective digital filters and algorithms is a fundamental aspect of modern signal processing, which finds applications across a wide spectrum of fields from audio processing to communications and biomedical engineering. Efficient and accurate digital filtering ensures high-quality performance and reliability in real-time applications.

A digital filter, in essence, is a mathematical algorithm that processes a discrete-time signal to reduce or enhance certain aspects of that signal. It can attenuate unwanted parts of the signal, such as noise, or enhance desired parts, such as certain frequencies. To implement a digital filter effectively, it is paramount to understand its design, structure, and the specific requirements of the application.

The design phase begins with identifying the type of filter needed. Digital filters are generally classified into two main categories: Finite Impulse Response (FIR) and Infinite Impulse Response (IIR) filters.

1. **Finite Impulse Response (FIR) Filters**: FIR filters have a finite duration response to an impulse input, meaning they settle to zero in a finite number of steps. Their primary advantage is that they can be designed to have exactly linear phase, meaning all frequency components of the input signal are shifted in time by the same amount. This characteristic is highly desirable in applications requiring phase-sensitive processing, such as data communications and audio signal processing. The design of an FIR filter usually involves specifying the filter's frequency response and using a windowing method or the Parks-McClellan algorithm to determine the filter coefficients.

2. **Infinite Impulse Response (IIR) Filters**: IIR filters, on the other hand, have an impulse response that, theoretically, goes on infinitely. They typically require fewer coefficients than FIR filters to achieve a similar level of performance, which can make them more computationally efficient. However, IIR filters are not inherently free from phase distortion unless they are compensated. The design of an IIR filter often involves analog filter

design methods followed by a transformation to the digital domain, using techniques such as the bilinear transform.

Once the type of filter is chosen, the implementation requires a careful balance between several factors: computational efficiency, memory usage, and numerical stability. In real-time applications, such as audio or video streaming, the filter must operate within stringent time constraints, processing each sample quickly to ensure smooth and uninterrupted output.

Computational efficiency can be enhanced by choosing an appropriate algorithm and optimization techniques. For example, the implementation of FIR filters can take advantage of convolution operations that exploit the symmetry properties of filter coefficients, reducing the number of required multiplications and additions. Additionally, Fast Fourier Transform (FFT) based methods can also be used to accelerate the convolution-based methods, especially in cases with long filter lengths.

Memory usage is another critical consideration, as digital filters often require storage for past input and output samples. Efficient memory management ensures that the filter does not consume excessive resources, which is especially important in embedded systems with limited memory.

Numerical stability is paramount, particularly for IIR filters that are more prone to instability due to their feedback nature. Ensuring stability involves carefully designing the filter coefficients, often through the use of software tools and simulations to analyze the filter's frequency response and stability margins.

Testing and validation are crucial steps in the implementation process. The designed filter must be rigorously tested with various input signals to ensure it meets the desired specifications. This includes checking its frequency response, phase response, and transient behavior.

In field deployments, continuous monitoring and adaptation might be necessary. Environmental changes and variability in input characteristics could necessitate real-time adjustments to the filter parameters, which can be achieved using adaptive filtering techniques. Adaptive filters adjust their coefficients dynamically based on the characteristics of the incoming signal, making them suitable for applications like noise cancellation and echo suppression.

In summary, implementing effective digital filters and algorithms is a complex yet essential task that requires a deep understanding of both theoretical and practical aspects of digital signal processing. The success of these implementations hinges on thoughtful design choices, efficient algorithmic strategies, careful optimization, and rigorous testing to ensure high performance and reliability in various real-world applications.

6.2 The Role of Fourier Transforms in Signal Processing

Fourier Transforms play an indispensable role in signal processing, bridging the gap between time-domain and frequency-domain analyses. By converting time-based signals into their frequency components, the Fourier Transform enables a clearer understanding of the underlying frequencies that build a signal, thereby facilitating more effective monitoring, filtering, and manipulation of these signals.

Signal processing encompasses various tasks, such as filtering noise from signals, compressing data for transmission, and enhancing signal resolution. These tasks often require a comprehensive representation of the signal's frequency content, which is provided by the Fourier Transform. Whether the application is in audio engineering, telecommunications, medical imaging, or radar systems, the ability to decompose a signal into its constituent frequencies offers significant advantages.

The Fourier Transform works on the premise that any time-domain signal can be represented as an infinite sum of sinusoidal functions—each with their own amplitude, frequency, and phase. This perspective is not only mathematically elegant but also practical. By breaking down a complex signal into simpler sinusoidal components, you can isolate and manipulate each frequency element independently. This decomposition can be crucial for identifying dominant frequencies in a signal or for filtering out unwanted noise.

A core application of the Fourier Transform in signal processing is the spectral analysis of signals. Imagine a situation where you are analyzing the frequency content of a recorded piece of music. Simply listening to the music might give you a qualitative understanding, but applying the Fourier Transform can quantitatively reveal the specific frequencies present in the audio. This can help in tasks like equalizing audio tracks, where specific frequency bands are amplified or attenuated to achieve a desired sound quality.

In the realm of telecommunications, Fourier Transforms are used extensively for modulating and demodulating signals. Modulation schemes, like amplitude modulation (AM) and frequency modulation (FM), are inherently functions of the frequency domain. During transmission, a signal is often mixed with a carrier frequency to shift it into a higher frequency band, reducing the antenna size and increasing the transmission range. The Fourier Transform facilitates this conversion, making it easier to transmit signals over long distances without significant loss of quality.

Moreover, digital signal processing (DSP), which relies on discrete forms of Fourier Transforms such as the Discrete Fourier Transform (DFT) and its fast computational variant, the Fast Fourier Transform (FFT), underpins modern digital communication systems. The FFT, in particular, is a cornerstone algorithm due to its computational efficiency, enabling real-time processing of signals across diverse applications. For instance, in cellular networks, the FFT is integral to the implementation of Orthogonal Frequency Division Multiplexing (OFDM), a method that vastly improves bandwidth efficiency and signal robustness.

Medical imaging technologies—like Magnetic Resonance Imaging (MRI) and Computed Tomography (CT)—also exploit Fourier Transforms. In MRI, for example, the spatial representation of objects within the body is reconstructed by applying the Fourier Transform to the received signals, rendering detailed images that are invaluable for diagnosis. The precision and reliability of Fourier-based techniques make them crucial for producing high-resolution images.

Filtering is another critical area where Fourier Transforms prove valuable. Consider the challenge of removing noise from a signal; in the time domain, noise and signal can be indistinguishable. Transforming the noisy signal into the frequency domain typically reveals that noise is concentrated at specific frequency bands, allowing targeted filtering without distorting the essential part of the signal. After the noise is removed in the frequency domain, the inverse Fourier Transform is applied to convert the signal back into the time domain with reduced noise.

In summary, the Fourier Transform is foundational in signal processing, providing essential tools for analyzing, transmitting, and transforming signals in various domains. Its ability to move between time and frequency domains enables detailed spectral analysis, efficient data transmission, high-resolution imaging, and effective noise filtering, demonstrating its critical role in modern engineering and scientific applications.

6.3 Managing Digital Signal Processing Across Systems

In the contemporary landscape of technology, managing digital signal processing (DSP) across various systems constitutes a cornerstone of efficient and high-quality digital communications, media, and data processing. Digital signal processing, fundamentally, encompasses the manipulation of signals—such as audio, video, temperature, or radar signals—once they are converted from analog to digital formats. This manipulation ranges from filtering and enhancing to compressing and analyzing, depending on the application requirements. Effective management of DSP across multiple systems necessitates an intricate balance of hardware capabilities, software sophistication, and system integration.

A significant aspect of DSP management is the hardware involved. Digital signals need to be processed in real-time, requiring high-performance processors capable of handling large volumes of data swiftly and accurately. The architecture of these processors typically includes features optimized for mathematical computations fundamental to DSP, such as Fast Fourier Transforms (FFT) or matrix multiplications. Modern systems employ dedicated DSP chips or multipurpose Central Processing Units (CPUs) and Graphics Processing Units (GPUs) with DSP capabilities. The choice of hardware directly affects the system's throughput, latency, power consumption, and overall performance.

Moreover, software frameworks and algorithms play a pivotal role in DSP. The development of efficient algorithms is essential to manage and optimize DSP tasks. For instance, algorithms for noise reduction, echo cancellation, and signal compression need to be tailored to meet the specific requirements of each application. These algorithms must also be scalable and adaptable to different processing environments, from embedded systems to large-scale server farms. The use of standardized programming languages and libraries, such as C, Python, MATLAB, and TensorFlow, fosters compatibility and accelerates development cycles.

Integration across different systems presents another layer of complexity. Systems that employ DSP often need to interact with various other systems, creating a need for seamless integration. This includes interoperability between different DSP hardware and software, as well as communication protocols that ensure data integrity and low latency. Middlewares and APIs (Application Programming Interfaces) are often used to bridge gaps between disparate systems, ensuring that the digital signals are processed correctly irrespective of the underlying hardware/software disparities.

The management of DSP across systems also demands rigorous consideration of latency and synchronization issues. In real-time applications such as online gaming, video conferencing, and autonomous vehicle navigation, signals must be processed with minimal delay. Latency in DSP can lead to significant performance degradation. Therefore, techniques such as pipeline processing, parallel computing, and real-time operating systems (RTOS) are employed to meet stringent timing requirements. Synchronization mechanisms, such as clock synchronization in distributed systems, ensure that data from different sources is processed coherently and concurrently.

Another critical factor is the power efficiency of DSP systems, particularly in battery-operated devices like smartphones, wearable technology, and IoT devices. These systems require energy-efficient DSP techniques to prolong battery life while maintaining high performance. Power management strategies, including dynamic voltage and frequency scaling (DVFS) and power gating, are frequently adopted in DSP hardware to optimize power consumption without compromising processing capabilities.

Furthermore, security in DSP management cannot be overlooked. As digital signals often carry sensitive information, protecting this data from unauthorized access and tampering is imperative. Encryption techniques safeguard the data during processing, ensuring confidentiality and integrity. Secure boot processes, hardware security modules (HSM), and secure communication protocols are deployed to create robust security frameworks around DSP systems.

Lastly, the future of managing DSP across systems is shaped by emerging technologies such as machine learning and artificial intelligence, which introduce novel approaches to signal processing. AI-driven DSP can dynamically adapt to changing signal characteristics and environments, offering unprecedented levels of efficiency and accuracy. Edge computing also plays a role, bringing DSP closer to the data source, thereby reducing latency and bandwidth usage. As these technologies evolve, the strategies for managing DSP will continue to innovate, pushing the boundaries of what is possible in digital signal processing.

In conclusion, managing digital signal processing across systems is a multi-faceted task that involves a symbiosis of hardware performance, software sophistication, system integration, synchronization, power efficiency, security, and emerging technologies. Each of these components must be meticulously engineered and harmonized to achieve optimal DSP management, facilitating superior performance and functionality in various applications.

6.4 Tools and Techniques for Improving Digital Signal Processing

Digital Signal Processing (DSP) is fundamental in various fields such as telecommunications, audio engineering, and biomedical engineering. With the increasing demand for high-quality signal processing applications, the development and utilization of improved tools and techniques are critical. These tools and techniques can enhance the functionality, efficiency, and accuracy of DSP systems. The following discussion covers several imperative tools and methodologies for improving DSP.

1. Fast Fourier Transform (FFT)

FFT is a crucial algorithm that efficiently computes the Discrete Fourier Transform (DFT) and its inverse. Traditional DFT operations are computationally intensive, rendering them impractical for real-time applications. FFT, by contrast, reduces the number of required computations dramatically—from $O(N^2)$ to $O(N \log N)$—where N is the number of data points. This substantial reduction in computational complexity makes FFT highly suitable for analyzing frequencies within a signal, enabling real-time processing for applications like audio filtering, image processing, and telecommunications.

2. Adaptive Filtering

Adaptive filtering algorithms dynamically adjust their parameters based on the input signal characteristics. These filters, including Least Mean Squares (LMS) and Recursive Least Squares (RLS), are particularly useful in environments where signal properties continuously evolve. For instance, in noise cancellation systems, adaptive filters can suppress unwanted noise while preserving the desired signal. The flexibility of these algorithms enables adaptive filters to outperform static filters in dynamic, real-world applications.

3. Multirate Signal Processing

Multirate signal processing involves using different sampling rates within a system, which helps optimize processing resources and improve system performance. By strategically varying the sampling frequency, it is possible to reduce the amount of data that needs processing, thereby lowering computational costs. Various functions like decimation and interpolation play a role in multirate signal processing. Decimation reduces the sampling rate by deleting samples, while interpolation increases it by adding samples. This technique is particularly useful in applications like compressing high-bandwidth signals or processing signals with diverse frequency components.

4. Wavelet Transform

Wavelet Transform (WT) is a powerful tool for analyzing localized variations of power within a signal. Unlike Fourier Transform, which provides frequency domains, WT offers both time and frequency representation. This makes it exceedingly valuable for applications such as image compression, seismic waveform analysis, and biomedical signal processing. WT can decompose a signal into components with scale-varying properties, making it easier to analyze transient features and irregular patterns within the signal.

5. Digital Filter Design

Digital filters are used to remove unwanted components from a signal, such as noise, or to extract useful parts of the signal. Advanced design techniques for digital filters include Finite Impulse Response (FIR) and Infinite Impulse Response (IIR) filters. FIR filters are inherently stable and simple to design, but they often require a higher order than IIR filters to achieve the desired response. IIR filters, on the other hand, can achieve a desired frequency response with fewer coefficients but may suffer from stability issues. Modern tools and algorithms significantly simplify the filter design process, enabling the creation of highly efficient and stable filters tailored to specific applications.

6. Machine Learning for DSP

The integration of machine learning (ML) techniques into DSP opens up new paradigms for the development and improvement of signal processing systems. ML algorithms can be trained to identify patterns and make decisions based on data, enabling more sophisticated signal enhancement, classification, and compression techniques. For instance, convolutional neural networks (CNNs) have been employed in audio signal processing to improve speech recognition and noise reduction systems substantially. Reinforcement learning models can adaptively optimize filter parameters for real-time applications, learning and improving performance on the fly.

7. FPGA and DSP Processors

Field Programmable Gate Arrays (FPGAs) and specialized DSP processors offer hardware-level improvements that significantly enhance processing speed and efficiency. FPGAs provide the advantage of reconfigurability, allowing for the implementation of custom DSP algorithms directly in hardware, leading to faster execution times and lower power consumption. DSP processors are optimized for complex mathematical operations essential in signal processing tasks, including multiply-accumulate (MAC) operations. Utilizing these hardware solutions can yield substantial improvements in processing throughput and system performance.

In conclusion, the advancement of tools and techniques for digital signal processing continues to evolve, driven by both theoretical advancements and practical requirements. Techniques such as FFT, adaptive filtering, multirate processing, and the integration of machine learning promise significant enhancements in DSP capabilities. Additionally, advancements in hardware solutions like FPGAs and dedicated DSP processors offer substantial improvements in efficiency and performance. By leveraging these tools and techniques, designers and engineers can create robust, efficient, and highly effective digital signal processing systems suitable for a broad range of applications.

7. Analog Signal Processing Techniques

In the ever-evolving landscape of technology, the seamless integration of digital and analog domains is crucial for achieving optimal performance in various applications. While digital signal processing has gained prominence due to its precision and versatility, analog signal processing techniques continue to hold a vital place, especially in scenarios where real-time processing, low latency, and high-frequency response are paramount.

Understanding and leveraging analog signal processing methods is not just a journey into the past but a necessary step towards enhancing contemporary systems. This chapter delves into the foundational concepts, practical applications, and intricate details that encompass the realm of analog signal processing. From filters and amplifiers to modulators and demodulators, it explores the diverse toolkit available to engineers and technologists aiming to harness the power of analog signals.

By examining case studies and practical examples, the chapter aims to provide a comprehensive understanding of how analog techniques can be effectively implemented and optimized. Whether you're an engineer seeking to deepen your technical expertise or a student embarking on the exploration of signal processing, this chapter serves as a crucial resource, bridging theoretical knowledge with practical application.

7.1 Enhancing Analog Signals with Filters and Modulation

In the realm of analog signal processing, the enhancement of signals is crucial for achieving clarity and reducing noise. This process relies heavily on two fundamental techniques: the use of filters and the application of modulation. Both elements play distinct yet complementary roles in refining the quality and reliability of analog signals for various applications such as communications, audio engineering, and instrumentation.

Filters are essential devices used to selectively allow or suppress certain frequencies of an analog signal. They can be broadly categorized into four types: low-pass, high-pass, band-pass, and band-stop filters. Each type serves a specific purpose in signal enhancement.

Low-pass filters are designed to allow signals with a frequency lower than a certain cutoff frequency to pass through while attenuating frequencies higher than the cutoff. These filters are especially useful in applications where high-frequency noise needs to be eliminated from a signal, such as in audio processing where it is vital to remove unwanted high-frequency hiss while retaining the essence of the sound.

Conversely, high-pass filters allow frequencies higher than a specific cutoff frequency to pass, attenuating those below this threshold. This type of filter is utilized in instances where low-frequency noise or drift needs to be removed. For example, in telecommunications, high-pass filters can block low-frequency components such as DC offsets that may interfere with the integrity of the transmitted signal.

Band-pass filters permit frequencies within a certain range to pass through while attenuating frequencies outside this range. These are particularly useful in applications like radio communications, where it is important to isolate a specific band of frequencies from a broad spectrum. By doing so, band-pass filters enable the reception of desired signals while minimizing the impact of adjacent channel interference.

Band-stop filters, or notch filters, do the opposite of band-pass filters by attenuating a specific range of frequencies and allowing frequencies outside of this range to pass through. These filters can be used to eliminate specific unwanted frequencies, such as the 60 Hz hum from power lines, without affecting the rest of the signal spectrum.

While filters serve to refine the frequency components of an analog signal, modulation is the technique used to encode information onto a carrier signal for transmission over various media. Modulation schemes can be divided into three main categories: amplitude modulation (AM), frequency modulation (FM), and phase modulation (PM).

In amplitude modulation, the amplitude of the carrier signal is varied in proportion to the information signal. This technique is widely used in broadcasting, such as AM radio, where it enables the transmission of audio signals over long distances. However, AM is susceptible to noise and interference, as any fluctuations in amplitude can distort the transmitted signal.

Frequency modulation addresses some of the shortcomings of AM by varying the frequency of the carrier signal in accordance with the information signal. FM is commonly used in radio broadcasting and two-way radio communications due to its resilience to amplitude-based noise and interference. The advantage of FM lies in its ability to provide high-fidelity audio transmissions, making it the preferred method for many high-quality broadcasting applications.

Phase modulation involves varying the phase of the carrier signal in response to the information signal. This technique is less common in analog applications but serves as the foundation for more complex digital modulation schemes. Phase modulation is inherently robust against certain types of noise and can be used in applications where phase integrity is critical.

Combining filters and modulation techniques enables the optimal enhancement and transmission of analog signals. Filters can be used to precondition a signal before modulation, ensuring that unwanted frequencies are removed and the desired signal components are preserved. Once modulated, the signal can then be transmitted with greater clarity and stability, reaching its destination with minimal interference.

In conclusion, the enhancement of analog signals through filters and modulation forms the backbone of various modern communication systems. By carefully selecting and applying these techniques, engineers can significantly improve the quality and reliability of analog signal transmission, ensuring that the intended information is accurately conveyed across different media. Thus, understanding and implementing these principles is essential for advancing the field of signal processing and communication technology.

7.2 The Importance of Noise Reduction Techniques

Noise reduction techniques play a pivotal role in a variety of fields, ranging from telecommunications and data processing to medical imaging and environmental science. The cornerstone of these techniques is the enhancement of signal clarity by minimizing unwanted noise, which can obscure or distort the information being conveyed.

One of the foremost reasons for the significance of noise reduction techniques is their capacity to enhance communication systems. In wireless communication, for example, noise—from sources such as other wireless devices, electrical equipment, or even atmospheric conditions—can interfere with signal transmission, leading to errors and reduced quality. Techniques like filtering, error correction, and advanced modulation schemes are employed to mitigate this noise, thereby improving the reliability and clarity of the information being transmitted. This reliability is crucial for applications ranging from everyday mobile phone conversations to critical emergency response communications.

In the realm of data processing, noise can originate from both external and internal sources, such as measurement errors, sensor inaccuracies, or even inherent randomness in the data itself. Techniques like data smoothing, outlier detection, and the use of algorithms designed to identify and remove anomalies are integral to ensuring data integrity. The significance here lies in the fact that accurate data is essential for decision-making processes, from business analytics to scientific research. Without effective noise reduction, insights drawn from the data could be flawed, leading to potentially costly or harmful decisions.

Medical imaging is another field where noise reduction techniques are of utmost importance. Imaging modalities such as MRI, CT scans, and X-rays are susceptible to various forms of noise, which can obscure the details necessary for accurate diagnosis. Techniques like spatial filtering, adaptive algorithms, and advanced reconstruction methods are employed to enhance image quality. Reduced noise in medical images facilitates better interpretation by healthcare professionals, leading to more accurate diagnoses and improved patient outcomes. Additionally, noise reduction can enable the use of lower doses of radiation in imaging techniques like CT scans, minimizing patient exposure to harmful effects.

Environmental science also relies heavily on noise reduction techniques, particularly in the collection and analysis of data related to air and water quality, climate change, and seismic activity. Sensors and measurement devices used in these fields often encounter noise from environmental factors or equipment limitations. Effective noise reduction ensures that the data collected is accurate, which is essential for monitoring environmental changes, predicting natural disasters, and implementing policy decisions related to environmental protection.

In the digital age, noise reduction is equally crucial in the realm of multimedia, including audio and visual content. For instance, in audio recording and broadcasting, noise can detract from the listener's experience by introducing unwanted sounds or distorting the original signal. Techniques such as noise gating, spectral subtraction, and adaptive filtering help to preserve the integrity of the audio signal, ensuring a clearer and more enjoyable listening experience. Similarly, in video production, noise can manifest as visual distortions or graininess, particularly in low-light conditions. Techniques like temporal filtering, spatial noise reduction, and machine learning-based approaches are utilized to enhance video quality, providing a clearer and more aesthetically pleasing visual output.

Furthermore, noise reduction techniques are imperative in the field of robotics and automation, where sensors and communication systems are prone to interference and inaccuracies. Effective noise reduction allows for more precise control and better decision-making by robotic systems, enhancing their performance and reliability in tasks ranging from industrial automation to autonomous vehicles.

In summary, the importance of noise reduction techniques cannot be overstated. They serve as fundamental tools across diverse applications, ensuring the accuracy and clarity of information, which in turn enhances decision-making, improves outcomes, and facilitates advancements in technology and science. Whether in communication systems, data processing, medical imaging, environmental monitoring, multimedia, or robotics, the efficacy of noise reduction techniques directly impacts the effectiveness and reliability of these systems, underscoring their critical role in modern society.

7.3 Managing Challenges in Analog Signal Processing

Analog signal processing is an essential domain within electrical engineering, integral to numerous applications, from audio and video processing to telecommunications and instrumentation. The inherent nature of analog signals, which are continuous in time and can take an infinite number of values, presents both opportunities and challenges. Effective management of these challenges requires a deep understanding of analog components, signal behavior, and the impact of various external and internal factors.

One primary challenge in analog signal processing is noise. Noise can originate from multiple sources such as thermal agitation of electrons within resistors (thermal noise), shot noise in transistors, and electromagnetic interference from external sources. Unlike digital signals, where it is relatively straightforward to distinguish between high and low states, analog signals can be significantly affected by even minor noise disturbances. Engineers must employ strategies such as shielding, filtering, and proper grounding to mitigate noise and preserve signal integrity.

Another critical challenge is distortion, which alters the original signal's waveform. Distortion can arise from non-linearities in amplifiers, mixers, and other signal processing elements. Harmonic distortion, where signals at multiples of the original frequency are introduced, can significantly degrade signal quality, particularly in high-fidelity audio systems. Engineers must carefully design and choose components with minimal distortion characteristics and employ feedback techniques to linearize the response of circuits.

Frequency response and bandwidth limitations also pose significant challenges. Every analog system has a finite bandwidth, beyond which the signal cannot be accurately processed. For instance, operational amplifiers and other analog components have a limited frequency response, affecting their performance at high frequencies. To manage this, engineers need to design circuits that adequately balance bandwidth, gain, and stability. Techniques such as compensation are used to extend the usable frequency range of amplifiers, ensuring reliable operation within the specified bandwidth.

Analog signals are also susceptible to drift and offset errors, particularly in precision applications. Drift can result from temperature variations, aging of components, and power supply fluctuations. Offset errors, where a constant deviation is present in the signal, can compromise the accuracy of measurements and control systems. Precision voltage references, temperature compensation techniques, and regular calibration are essential measures to control drift and offset errors.

Phase and amplitude errors can further complicate the design and performance of analog systems. These errors can skew signal timing and amplitude, leading to synchronization issues and inaccurate signal interpretation. Phase-locked loops (PLLs) and automatic gain control (AGC) circuits are often deployed to correct phase and amplitude discrepancies, ensuring that the processed signal maintains fidelity with the original input.

Impedance matching is another vital aspect of analog signal processing. Mismatches between source and load impedances can cause reflections, standing waves, and signal loss, particularly in high-frequency applications. Engineers must design circuits that properly match impedances to ensure maximum power transfer and minimize signal degradation. Techniques like using impedance matching networks, transformers, and adaptive circuits are employed to achieve this.

Integration and miniaturization of analog circuits bring their unique set of challenges, especially in modern electronic devices. Packing numerous components in limited space can lead to crosstalk, where signals in one circuit inadvertently affect another. Proper layout design, the use of ground planes, and isolation techniques are critical to reducing crosstalk and ensuring reliable operation of densely packed analog circuits.

Power supply management is crucial in analog signal processing. Analog circuits are often sensitive to power supply fluctuations and noise, which can directly influence signal integrity. Engineers employ low-noise voltage regulators, power supply filters, and decoupling capacitors to maintain clean and stable power sources for analog circuits.

Lastly, the analog-digital interface poses significant considerations. Converting analog signals to digital (and vice versa) must be handled with precision to maintain signal integrity. Analog-to-digital converters (ADCs) and digital-to-analog converters (DACs) must be chosen carefully, considering factors such as resolution, sampling rate, and linearity. Anti-aliasing filters are necessary to prevent high-frequency noise from corrupting the conversion process.

Managing challenges in analog signal processing is a multifaceted endeavor that requires a thorough understanding of various physical and technical phenomena. By effectively addressing issues such as noise, distortion, bandwidth limitations, and impedance matching, engineers can design robust analog circuits that perform reliably across a wide range of applications.

7.4 Case Studies of Effective Analog Signal Processing

Analog signal processing has stood the test of time as an essential technique in various domains, including telecommunications, audio engineering, medical instrumentation, and radars. Understanding real-world applications through case studies elucidates how analog signal processing is effectively implemented to solve practical problems. This section delves into seven notable case studies that showcase the diverse and efficient applications of analog signal processing.

Case Study 1: Telecommunications Filtering Systems

One of the foundational applications of analog signal processing is in the design of filtering systems for telecommunications. A quintessential example is the use of band-pass filters to isolate specific frequency bands in radio transmission. These filters allow communication systems to eliminate unwanted signals and interference, thus ensuring clearer and more reliable communication. By implementing precise analog filters, engineers can optimize the signal-to-noise ratio, leading to higher fidelity in voice and data transmission.

Case Study 2: Audio Engineering and Sound Quality

Analog signal processing plays a vital role in audio engineering, particularly in mixing and mastering audio tracks. Analog equalizers, compressors, and other signal processing devices are commonly employed to enhance sound quality. For instance, a renowned music producer may utilize an analog equalizer to balance the frequencies of a track, emphasizing the bassline while attenuating harsh high frequencies. This meticulous tweaking can significantly enrich the auditory experience, contributing to the production of high-quality audio content.

Case Study 3: Medical Instruments and Physiological Signal Processing

In the medical field, analog signal processing is crucial for the accurate interpretation of physiological signals. Electrocardiograms (ECGs) and electroencephalograms (EEGs) are classic examples where analog filters are used to extract meaningful information from biopotential signals. By filtering out noise and artifacts, these systems can provide clear representations of heart and brain activities, respectively. This enables healthcare professionals to make precise diagnoses and monitor patient conditions effectively.

Case Study 4: Radar and Sonar Systems

Radar and sonar systems rely heavily on analog signal processing to detect and locate objects. These systems emit signals and process the returned echoes to determine the range, speed, and other characteristics of targets. Through the use of analog mixers, filters, and amplifiers, radar systems can discern between closely spaced objects and minimize the impact of noise. Such precise signal processing allows for the accurate tracking of aircraft, ships, and even weather patterns.

Case Study 5: Analog Data Acquisition

In scientific research, data acquisition systems often use analog signal processing to capture and analyze experimental data. For example, in a physics laboratory studying electromagnetic phenomena, analog-to-digital converters (ADCs) are used to digitize analog signals from sensors. Before conversion, analog signal conditioning circuits might be employed to amplify and filter the signals, ensuring that only relevant information is captured. This careful preprocessing enhances the accuracy and reliability of the data collected.

Case Study 6: Seismic Exploration and Earthquake Detection

Seismic exploration for oil and gas as well as earthquake detection harness analog signal processing to identify and interpret seismic waves. Analog seismometers record ground vibrations, which are then processed to differentiate between natural seismic activity and potential hydrocarbon reservoirs. Analog filters are essential in isolating relevant frequencies and reducing noise, thereby improving the resolution and accuracy of the seismic images.

Case Study 7: Consumer Electronics and Analog Synthesisers

In the realm of consumer electronics, analog signal processing is ubiquitous, particularly in synthesisers used by musicians. Analog synths generate and manipulate sound in real-time through oscillators, filters, and modulators. Musicians often prefer analog synthesizers due to their warm, rich sound, which is a result of the continuous nature of analog signals.

These instruments allow for a high degree of control over the sound, enabling artists to create unique and expressive musical compositions.

In conclusion, analog signal processing remains a critical component across various fields, providing unparalleled precision and reliability in real-time applications. These case studies highlight the versatility and effectiveness of analog techniques in solving complex problems and advancing technology. Whether in telecommunications, audio engineering, medical instrumentation, or other areas, analog signal processing continues to play an indispensable role in modern technology.

8. The Role of Hardware in Signal Processing

In the intricate world of signal processing, the indispensable role of hardware cannot be overstated. While algorithms and software provide the theoretical foundations and flexibility needed for manipulating and interpreting signals, it is the hardware that often dictates the practical capabilities and limitations of these processes. As we delve deeper into this critical aspect of signal processing, we will explore the various types of hardware used, their specific functions, and how they interface with the software to achieve seamless signal manipulation.

Understanding the interplay between hardware and software in signal processing allows us to fully appreciate the performance enhancements and real-time capabilities that modern systems offer. From analog-to-digital converters (ADCs) and digital signal processors (DSPs) to field-programmable gate arrays (FPGAs) and application-specific integrated circuits (ASICs), each piece of hardware plays a pivotal role. These components not only determine the speed and efficiency of signal processing tasks but also influence factors such as power consumption, size, and overall system cost.

In this chapter, we will begin by examining the fundamental hardware components and their respective functions within signal processing systems. Following that, we will discuss the criteria for selecting appropriate hardware based on specific application needs, taking into account factors such as processing power, latency, and resource constraints. We will also touch upon emerging hardware technologies and their potential impact on future signal processing advancements.

Ultimately, a comprehensive understanding of the role of hardware in signal processing empowers engineers and developers to design more efficient, reliable, and high-performance systems, paving the way for innovative applications across various industries. As we embark on this journey, let us uncover the myriad ways in which hardware underpins and enhances the art and science of signal processing.

8.1 Leveraging DSP Chips for Processing Success

Digital Signal Processing (DSP) chips have become indispensable in modern technology, revolutionizing the way data is processed and interpreted across various applications. Leveraging these specialized microprocessors for processing tasks offers significant advantages in terms of efficiency, performance, and flexibility. These benefits arise from the unique architecture and functionality of DSP chips, making them ideal for a plethora of signal processing tasks.

At the core of DSP chips lies their ability to perform mathematical computations very rapidly. Unlike general-purpose processors (GPUs), DSP chips are specifically designed to handle complex mathematical operations such as multiplication and addition, which are fundamental to signal processing. This targeted design allows them to achieve higher performance speeds in signal processing tasks, which is crucial for applications requiring real-time data analysis.

One of the primary applications of DSP chips is in telecommunications. With the ever-growing demand for faster and more reliable communication, DSP chips are utilized to process signals in devices like mobile phones, base stations, and satellite communications. They handle functions such as signal compression, filtering, and modulation, which are essential for transmitting large amounts of data efficiently and accurately. For instance, the compression algorithms enabled by DSP chips allow for the reduction of data size, facilitating faster transmission rates without compromising the quality of the signal.

In addition to telecommunications, DSP chips play a crucial role in audio processing. From portable music players to professional audio equipment, these chips are behind the scenes ensuring high-quality audio reproduction and recording. DSP chips are used to implement sophisticated audio effects, noise reduction algorithms, and equalization processes. Their capability to rapidly execute these tasks allows for real-time audio processing, which is necessary for both live performances and studio recordings. By leveraging DSP technology, devices can deliver superior audio clarity and immersive sound experiences.

Medical technology also benefits significantly from the utilization of DSP chips. In diagnostic equipment like MRI and ultrasound machines, DSP chips are employed to enhance image quality and processing speed. They enable the real-time analysis of complex medical data, which is crucial for accurate diagnosis and treatment planning. For instance, in ultrasound imaging, DSP chips are responsible for processing the reflected signals to produce clear and detailed images of internal body structures. This rapid processing power facilitates timely medical decisions and improves patient outcomes.

The automotive industry harnesses the capabilities of DSP chips in the development of advanced driver-assistance systems (ADAS) and autonomous vehicles. DSP chips process data from various sensors, including cameras, lidar, and radar, to interpret the vehicle's surroundings. This processing power allows for real-time decision-making, essential for the safe and efficient operation of autonomous driving systems. Furthermore, DSP chips contribute to the development of in-car entertainment systems, providing superior audio and multimedia experiences.

In the realm of consumer electronics, DSP chips are integrated into devices like smartphones, digital cameras, and televisions to enhance functionality and performance. For instance, in smartphones, these chips are used for everything from improving call quality to enhancing camera image processing. In televisions, DSP chips facilitate high-definition video playback and upscale content from lower resolutions, providing viewers with a better visual experience. Their ability to handle intensive processing tasks without draining significant power also makes them ideal for portable electronic devices where battery life is a concern.

The adaptability of DSP chips is another significant advantage. Their programmable nature allows engineers to update and optimize algorithms as processing requirements evolve, without the need for hardware changes. This flexibility is particularly beneficial in rapidly changing technological landscapes, where the ability to quickly adapt and incorporate new functionality can provide a competitive edge.

In conclusion, leveraging DSP chips for processing offers numerous advantages across various fields, including telecommunications, audio processing, medical technology, automotive systems, and consumer electronics. Their specialized architecture enables rapid execution of complex mathematical operations, leading to enhanced performance, efficiency, and flexibility. As technology continues to evolve, the role of DSP chips is likely to expand further, driving innovations and improvements in data processing and interpretation across multiple industries. Embracing these powerful tools can lead to significant advancements and successes in diverse applications, making DSP chips a cornerstone of modern technology.

8.2 The Impact of FPGA in Signal Processing

Field-Programmable Gate Arrays (FPGAs) have revolutionized the landscape of signal processing by providing remarkable flexibility, performance, and efficiency. The impact of FPGAs in this domain can be explored through various aspects, including their architectural advantages, customization capabilities, performance benefits, and influence on innovation in signal processing applications.

The architectural structure of FPGAs provides a unique blend of hardware density and programmability. Unlike traditional microprocessors or ASICs (Application-Specific Integrated Circuits), FPGAs comprise an array of programmable logic blocks interconnected by a reconfigurable network. This structure enables parallel processing, which is highly beneficial for signal processing tasks that typically involve high data rates and complex mathematical operations. The inherent parallelism of FPGAs allows them to execute multiple signal processing functions simultaneously, reducing latency and increasing throughput.

Customization is another significant advantage of FPGAs in signal processing. Engineers can tailor the FPGA's configuration to meet specific application requirements, optimizing the hardware for the precise mathematical operations prevalent in signal processing. For example, tasks such as Fast Fourier Transform (FFT), filtering, and modulation can be mapped onto an FPGA in a way that maximizes the use of resources while minimizing power consumption and computation time. This flexibility is not commonly found in fixed-function hardware solutions, making FPGAs an attractive option for a broad range of signal processing applications.

Performance is paramount in signal processing, and FPGAs excel in this regard. They offer the capability to handle high-bandwidth data streams efficiently, making them ideal for real-time processing needs in applications such as telecommunications, radar systems, biomedical imaging, and audio/video processing. The ability to process data in real-time ensures that critical information is analyzed and acted upon without delay, which is crucial in applications where timing and speed are of the essence. FPGAs also support high-precision arithmetic operations, which are essential for maintaining accuracy in complex signal processing algorithms.

Another impact of FPGAs in signal processing is their influence on innovation. The continuous advancements in FPGA technology, including increased logic density, faster I/O capabilities, and integrated DSP (Digital Signal Processing) slices, have enabled researchers and engineers to push the boundaries of what is possible in signal processing. For instance, modern FPGAs incorporate specialized hardware blocks specifically designed to accelerate common DSP functions, reducing the development time and effort required to implement sophisticated algorithms. This feature has opened up new possibilities in areas like adaptive filtering, machine learning in signal processing, and advanced modulation schemes.

Moreover, FPGAs foster a rapid prototyping environment. When developing new signal processing algorithms, the ability to quickly iterate and test on FPGA hardware accelerates the research and development cycle. Designers can implement, verify, and refine their algorithms on the same hardware platform that will be used in the final product, ensuring a smoother transition from prototype to production. This capability drastically reduces time-to-market for new signal processing technologies.

The cost-effectiveness of FPGAs also cannot be overlooked. While the initial investment in FPGA development tools and expertise may be high, the long-term benefits often outweigh these costs. FPGAs provide a reconfigurable solution that can be updated and repurposed as standards evolve or application requirements change, extending the lifecycle of the hardware. This adaptability reduces the need for frequent hardware replacements, leading to significant savings, especially in industries where signal processing technology evolves rapidly.

In conclusion, the impact of FPGAs in signal processing is profound and multifaceted. Their ability to provide high performance, customization, flexibility, and cost-effectiveness makes them indispensable in modern signal processing applications. As FPGA technology continues to advance, its role in enabling new innovations and improving existing systems is likely to grow even more significant, driving further advancements in the field of signal processing.

8.3 Using Microcontrollers and Processors Effectively

Microcontrollers and processors serve as the brains of numerous technological devices, from household appliances to sophisticated industrial machinery. Understanding how to utilize these essential components effectively is crucial for any engineer or developer. With the advent of the Internet of Things (IoT) and increasing automation, the effective use of microcontrollers and processors has become more pertinent than ever.

Fundamentally, a microcontroller is a compact integrated circuit designed to govern a specific operation in an embedded system. It typically combines a processor, memory, and input/output peripherals on a single chip. Processors, on the other hand, are more versatile and can be found at the heart of almost every computing system. They execute programmed instructions to perform a variety of tasks. While microcontrollers are employed in specific, repetitive control applications, processors are utilized for general-purpose computing.

To use these components effectively, the first step involves selecting the right microcontroller or processor for the task. Several factors need consideration, such as computational requirements, power consumption, memory needs, and the nature of input/output interfaces. For instance, an application requiring basic control functions might be well-served by an 8-bit microcontroller, whereas more complex tasks like image processing could require a 32-bit processor.

Once the appropriate component is selected, the next step is to program the device efficiently. Writing optimized code is imperative to ensure that the microcontroller or processor runs smoothly within its operational constraints. This requires a deep understanding of the architecture of the selected component, including its instruction set, memory organization, and peripheral interfaces.

Effective use of interrupts is another critical aspect of programming these devices. Interrupts allow the processor or microcontroller to respond promptly to external events, thereby improving the system's responsiveness and multitasking abilities. However, careless use of interrupts can lead to issues such as priority inversion or interrupt latency. Therefore, it is essential to prioritize and manage interrupts carefully.

Power management is another important consideration, especially in battery-operated or energy-constrained applications. Techniques like sleep modes, power gating, and dynamic voltage scaling can significantly reduce power consumption. Choosing components with low-power features and optimizing the code to make the best use of these features can extend the battery life of the device.

Memory management is equally crucial. Microcontrollers often have limited RAM and flash memory, so it is essential to utilize these resources judiciously. Efficient memory management involves minimizing the use of global variables, using memory-efficient algorithms, and leveraging techniques like loop unrolling and inlining wherever appropriate. Additionally, effective use of direct memory access (DMA) can offload the processor and improve overall system performance.

Peripheral integration is another area where developers need to focus. Many microcontrollers come with built-in peripherals such as ADCs, DACs, timers, and communication interfaces (UART, I2C, SPI, etc.). Understanding how to configure and use these peripherals can greatly enhance the functionality of the system while freeing up processing power for other tasks.

Security is another essential aspect, particularly in IoT applications where microcontrollers often operate in a networked environment. Techniques such as secure boot, encryption, and authentication should be implemented to protect the system from malicious attacks. Using hardware features like Trusted Execution Environments (TEE) and secure elements can further bolster security.

In addition to the technical aspects, effective project management practices can play a significant role in the successful deployment of microcontroller-based systems. Following a structured development process, from requirement analysis to design, implementation, testing, and maintenance, can ensure that the project meets its objectives on time and within budget.

Finally, continuous learning and staying updated with the latest developments in microcontroller and processor technologies are essential.

The field is constantly evolving, and new architectures, programming techniques, and development tools regularly emerge. Engaging with the community, attending workshops and conferences, and reading relevant research papers can provide valuable insights and keep one ahead in the game.

In conclusion, the effective use of microcontrollers and processors involves a careful selection process, efficient programming, diligent resource management, and staying abreast of technological advancements. Mastering these aspects can lead to the development of robust, efficient, and innovative systems across a wide range of applications.

8.4 The Future of Hardware in Signal Processing

As advancements in technology continue to accelerate, the field of signal processing is poised for significant transformations driven by the evolution of hardware. This section delves into the prospective innovations and trends that are likely to shape the future landscape of hardware used in signal processing.

One of the noteworthy trends is the progression in semiconductor technologies. Moore's Law, which posits that the number of transistors on a microchip doubles approximately every two years, has driven the development of increasingly powerful and compact integrated circuits. While some experts argue that we are reaching the physical limits of transistor miniaturization, others believe innovations such as 3D stacking of semiconductor devices and the development of new materials like graphene will continue to push the boundaries. These advancements promise to enhance the performance and energy efficiency of processors, which are crucial for handling the complex computations involved in signal processing.

Quantum computing presents another revolutionary paradigm that could dramatically alter signal processing hardware. Unlike classical computers, which use bits as the smallest unit of information, quantum computers use quantum bits or qubits. Qubits can exist in multiple states simultaneously due to the principles of superposition and entanglement. This property allows quantum computers to perform parallel computations at unprecedented speeds, which could accelerate processes such as Fourier transforms and filtering operations that are fundamental to signal processing. Although practical, large-scale quantum computers are still in development, their potential benefits make them a compelling area of research and investment.

Another technological frontier is the use of artificial intelligence (AI) and machine learning (ML) in hardware design. Custom AI accelerators like Google's Tensor Processing Unit (TPU) and Nvidia's Tensor Cores are specifically optimized for machine learning tasks and can significantly boost signal processing tasks when integrated with AI algorithms. These hardware components can be particularly effective in applications requiring real-time signal analysis, such as autonomous vehicles and advanced robotics, where the ability to quickly and accurately process sensory data is critical.

Moreover, advancements in neuromorphic computing—an approach that aims to mimic the structure and function of the human brain—are gaining traction. Neuromorphic chips, such as Intel's Loihi, are designed to implement neural networks directly in hardware, enabling efficient and high-speed processing of complex signal patterns. This approach not only enhances computational speed but also significantly reduces power consumption, making it ideal for battery-operated devices and large-scale data centers.

The evolution of wireless communication technologies, such as 5G and the forthcoming 6G, will also have a significant impact on signal processing hardware. These technologies promise higher data rates, lower latency, and improved connectivity, which will necessitate more sophisticated hardware to manage and process the vast amounts of data generated. This includes advancements in antennas, RF circuits, and digital signal processors (DSPs) that can handle these increased demands with minimal delay.

In terms of material science, the development of new substrates and components that can operate at higher frequencies and temperatures is another area with substantial potential. Emerging materials such as Gallium Nitride (GaN) and Silicon Carbide (SiC) offer superior thermal conductivity and electron mobility compared to traditional silicon. These properties make them suitable for high-power and high-frequency applications, including radar systems, satellite communications, and other critical signal processing functions.

Additionally, the integration of hardware and software continues to evolve, leading to more holistic approaches to system design. Field-Programmable Gate Arrays (FPGAs), for instance, offer flexibility by allowing designers to reconfigure hardware post-manufacture to meet specific signal processing requirements. This adaptability is particularly valuable in fields where standards and algorithms frequently evolve, such as telecommunications and multimedia.

Finally, as society increasingly prioritizes sustainability, green computing initiatives are focusing on developing energy-efficient hardware without compromising performance. This trend is essential in reducing the environmental impact of large-scale data processing and extends the operational lifespan of battery-powered devices.

In conclusion, the future of hardware in signal processing is enriched by a confluence of advancements across various technological domains. As semiconductor technologies continue to evolve, quantum computing, AI and ML integration, neuromorphic computing, advanced wireless communication technologies, material science innovations, and holistic hardware-software integration will collectively shape the next generation of signal processing capabilities. These innovations promise not only to enhance the efficiency and power of signal processing hardware but also to open new horizons for applications that were previously unimaginable.

9. Signal Processing in Different Fields

Signal processing, a cornerstone of modern technology, finds its application across a multitude of diverse fields, each harnessing its capabilities to address unique challenges and innovate continuously. From the highly specialized realm of medical imaging to the ubiquitous domain of telecommunications, the principles and techniques of signal processing shape the foundational infrastructure of our digital and analog worlds. This chapter aims to transcend the theoretical aspects and explore the real-world implications and advancements facilitated by signal processing. By delving into various sectors such as audio engineering, finance, radar systems, and beyond, we uncover the tangible impact and transformative potential of signal processing technologies. Insightful case studies and cutting-edge research will illuminate how distinct industries adapt and innovate using these robust techniques, offering a comprehensive understanding of the far-reaching influence and versatility of signal processing.

9.1 Signal Strategies in Telecommunications

In the modern telecommunications landscape, the emphasis on signal strategies plays a critical role in ensuring efficient and reliable communication. This complex field encompasses various techniques and methodologies designed to optimize the transmission and reception of data across diverse communication systems. Signal strategies are fundamental to both wired and wireless networks, addressing challenges such as signal degradation, interference, bandwidth limitations, and overall transmission efficiency.

One of the most fundamental aspects of signal strategies is modulation, a process that involves varying a carrier signal to encode the information being transmitted. Modulation schemes can be broadly categorized into analog and digital forms. Analog modulation techniques, such as Amplitude Modulation (AM) and Frequency Modulation (FM), have been traditionally used in radio and television broadcasting. Digital modulation methods, including Phase Shift Keying (PSK), Frequency Shift Keying (FSK), and Quadrature Amplitude Modulation (QAM), are more commonly employed in modern telecommunications systems due to their ability to efficiently handle high data rates and provide better noise immunity.

Error detection and correction are also integral components of signal strategies. These techniques are crucial for maintaining data integrity over noisy or unreliable communication channels. Error detection schemes, such as parity checks and cyclic redundancy checks (CRC), allow for the identification of errors in transmitted data. Correction techniques, like Forward Error Correction (FEC) methods including Reed-Solomon and convolutional codes, enable receivers to correct certain errors without requiring retransmission.

Another critical signal strategy is multiplexing, which maximizes the use of available bandwidth by combining multiple signals for transmission over a single communication channel. Time Division Multiplexing (TDM) and Frequency Division Multiplexing (FDM) are traditional techniques used in telephony and broadcasting. Contemporary methods, such as Code Division Multiple Access (CDMA) and Orthogonal Frequency Division

Multiplexing (OFDM), have been widely adopted in cellular networks and broadband internet services. These advanced multiplexing schemes allow for the simultaneous transmission of multiple signals without significant interference, enhancing both capacity and efficiency.

Signal processing techniques further enhance telecommunications by improving signal quality and extracting desired information from noisy or distorted data. Digital Signal Processing (DSP) algorithms are employed to perform various functions, including filtering, compression, and spectral analysis. DSP techniques are essential in applications such as voice and video compression, echo cancellation, and adaptive filtering, which help maintain signal clarity and reduce latency.

In wireless communication, dealing with multipath propagation—where signals take multiple paths to reach the receiver—is a significant challenge. Strategies such as diversity schemes and MIMO (Multiple Input Multiple Output) technology are employed to mitigate the adverse effects of multipath fading. Diversity schemes, including time, frequency, and spatial diversity, utilize multiple transmission and reception paths to increase the likelihood of receiving a clear signal. MIMO technology, which involves multiple antennas at both the transmitter and receiver, exploits spatial diversity to enhance signal quality and increase data throughput.

Spectrum management also plays an essential role in signal strategies, particularly in the increasingly crowded radio frequency environment. Regulatory agencies allocate and manage spectrum resources to prevent interference and ensure efficient utilization. Techniques such as dynamic spectrum access and cognitive radio enable more flexible and efficient use of the available spectrum. These approaches allow devices to dynamically adjust their operating frequencies based on real-time assessments of the spectrum environment, thereby reducing congestion and improving overall communication performance.

Furthermore, advances in signal encryption ensure secure communication, protecting data from unauthorized access and tampering. Cryptographic algorithms and protocols, such as AES (Advanced Encryption Standard) and RSA (Rivest-Shamir-Adleman), are integral to modern telecommunications, providing robust security mechanisms for data transmission.

In summary, the intricate web of signal strategies in telecommunications encompasses a wide range of techniques, from modulation and error correction to multiplexing, signal processing, and spectrum management. These strategies are vital for overcoming the challenges of signal degradation, interference, and limited bandwidth, thereby ensuring the efficiency, reliability, and security of communication systems that form the backbone of our interconnected world.

9.2 Signal Processing in Biomedical Engineering

In the field of biomedical engineering, signal processing is an essential tool that serves as the linchpin for acquiring, analyzing, and interpreting complex biological signals. The primary aim is to extract meaningful information from physiological data, which can then be used for diagnosis, monitoring, and treatment of medical conditions. The realm of biomedical signal processing encompasses a diverse range of biological signals, including but not limited to electrocardiograms (ECG), electroencephalograms (EEG), electromyograms (EMG), and biomedical imaging signals like MRI and CT scans.

Signal processing in biomedical engineering begins with the acquisition of raw signals. Devices such as electrodes, sensors, and imaging equipment capture the raw data from the human body. However, this data often contains noise and artifacts from various sources, including muscle contractions, power line interference, and electrode motion. Therefore, the initial phase of signal processing involves preprocessing steps such as filtering, signal enhancement, and artifact removal. Techniques such as low-pass, high-pass, band-pass, and notch filters are commonly employed to remove unwanted frequencies and enhance the quality of the signals.

Once the signals are preprocessed, the next step is feature extraction. Feature extraction aims to identify and isolate specific characteristics of the signal that are relevant for further analysis. For instance, in ECG signal processing, critical features such as the P-wave, QRS complex, and T-wave are identified to evaluate heart function. Time-domain analysis, where features such as amplitude, duration, and intervals are measured, provides vital information about the physiological state. In contrast, frequency-domain analysis employs transformations like the Fourier Transform to study the spectral components of the signal, enabling the detection of periodicities and abnormalities that are not evident in the time domain.

The extracted features often contribute to the construction of models for classification, diagnosis, or prediction. These models can be developed using various methodologies, such as statistical analysis, machine learning, and deep learning. In machine learning, classifiers like Support Vector Machines (SVM), k-Nearest Neighbors (k-NN), and decision trees are utilized to categorize signals into different classes, such as normal or pathological states. Deep learning techniques, particularly Convolutional Neural Networks (CNN) and Recurrent Neural Networks (RNN), have shown effective in handling large and complex datasets, offering superior performance and accuracy in tasks like automated diagnosis and image interpretation.

Signal processing also plays a crucial role in biomedical imaging. Techniques like Magnetic Resonance Imaging (MRI) and Computed Tomography (CT) rely heavily on sophisticated signal processing algorithms to reconstruct images from raw data. In MRI, raw data captured from the magnetic resonance signals is transformed using the Fourier Transform to create detailed images of internal body structures. Similarly, in CT imaging, signal processing algorithms reconstruct cross-sectional images from X-ray data, aiding in the visualization of tissues, organs, and abnormalities.

Another significant aspect of biomedical signal processing is the real-time processing ability, which is crucial for applications such as wearable health monitoring devices and implantable medical devices like pacemakers. These devices continuously collect and process signals to monitor vital parameters, providing immediate feedback and triggering alarms or therapeutic interventions when abnormal patterns are detected.

Advancements in signal processing techniques have also fostered the development of telemedicine and remote monitoring systems. These systems leverage wireless communication technologies to transmit processed biomedical data from patients to healthcare providers, enabling remote diagnosis, monitoring, and consultation, which is especially vital in remote or underserved regions.

In conclusion, signal processing is fundamental to biomedical engineering, bridging the gap between raw physiological data and actionable medical insights. It encompasses a broad spectrum of techniques and applications, from noise reduction and feature extraction to complex modeling and real-time monitoring, all aimed at improving patient outcomes and advancing medical science. As technology continues to evolve, the role of signal processing in biomedical engineering is poised to become even more integral, driving innovations in diagnostics, treatment, and healthcare delivery.

9.3 Managing Signal Processing in Consumer Electronics

Signal processing in consumer electronics is an integral field that deals with the manipulation, analysis, and interpretation of signals, such as sound, images, and other data. As consumer electronics have evolved, the complexity and efficiency of signal processing have become paramount in delivering high-quality products that meet the needs and expectations of consumers. Managing signal processing involves several essential components, techniques, and considerations that are necessary for the optimal functioning of these devices.

One of the key aspects of managing signal processing is the design and implementation of algorithms. Algorithms are mathematical procedures or formulas that perform operations on data to achieve a specific outcome. In consumer electronics, these algorithms can be used for a variety of purposes, including noise reduction, compression, enhancement, and feature extraction. For instance, in audio processing, algorithms are used to filter out background noise, adjust equalization settings, and implement compression techniques to ensure clear and balanced sound. Similarly, in image processing, algorithms help in reducing noise, enhancing resolution, and improving color accuracy.

Another critical aspect of managing signal processing is hardware optimization. This involves selecting and configuring the appropriate processors, memory, and other hardware components to efficiently execute signal processing algorithms. With the rapid advancement in semiconductor technology, modern consumer electronics are equipped with specialized hardware, such as Digital Signal Processors (DSPs) and Graphics Processing Units (GPUs), which are designed to handle the rigorous demands of signal processing tasks. These specialized processors offer higher computational power, energy efficiency, and real-time processing capabilities compared to general-purpose processors.

Software development is another pillar in the management of signal processing. The development of robust and efficient software is crucial for implementing signal processing algorithms on hardware platforms. This involves writing code in programming languages such as C, C++, and Python, which can efficiently utilize the hardware resources. Additionally, software development often involves the use of specialized libraries and frameworks that provide pre-designed functions and tools for signal processing tasks. These libraries, such as MATLAB, OpenCV, and TensorFlow, enable developers to quickly prototype, test, and optimize algorithms.

Energy efficiency is a significant consideration in managing signal processing for consumer electronics. With the increasing demand for portable and battery-operated devices, such as smartphones, tablets, and wearable technology, optimizing energy consumption has become a priority. Efficient power management techniques, such as dynamic voltage and frequency scaling (DVFS) and low-power design methodologies, are employed to reduce the energy consumption of signal processing operations without compromising performance. This ensures longer battery life and improved user experience.

Real-time processing is another essential aspect of signal processing management. Many consumer electronics applications, such as video streaming, gaming, and virtual reality, require real-time processing to deliver seamless and interactive experiences. Real-time processing necessitates the efficient handling of data with minimal latency and jitter. Techniques such as parallel processing, multi-threading, and hardware acceleration are often employed to meet real-time performance requirements.

Security and privacy are increasingly important factors in the management of signal processing. As consumer electronics become more interconnected and data-driven, safeguarding sensitive information and protecting against unauthorized access have become critical. Signal processing techniques, such as encryption, authentication, and watermarking, are implemented to ensure the security and integrity of data transmitted and processed by consumer electronic devices.

Finally, system integration and testing are crucial in managing signal processing. The integration of signal processing algorithms, hardware, and software components requires meticulous planning and coordination to ensure compatibility and optimal performance. Rigorous testing and validation processes are conducted to identify and rectify any issues or inefficiencies. This involves functional testing, performance testing, and user acceptance testing to ensure that the final product meets the specified requirements and delivers a high-quality user experience.

In conclusion, managing signal processing in consumer electronics is a multifaceted endeavor that encompasses algorithm design, hardware optimization, software development, energy efficiency, real-time processing, security, and system integration. As consumer electronics continue to advance, effective management of signal processing will remain a cornerstone in the development of innovative and high-performance products.

9.4 Signal Processing in Environmental Monitoring

Signal processing plays a critical role in environmental monitoring, serving as the backbone for accurate data acquisition, analysis, and interpretation. As our world grapples with climate change, pollution, and other ecological challenges, the sophisticated techniques of signal processing become vital for timely and precise environmental assessment.

At its core, signal processing involves the manipulation and analysis of signals to extract valuable information. In environmental monitoring, these signals often come from various sensors that detect phenomena such as temperature, humidity, atmospheric pressure, pollutant concentration, and seismic activity. The collected data typically contain noise and redundancies, necessitating robust signal processing techniques to filter out irrelevant information and highlight critical insights.

One fundamental aspect of signal processing in environmental monitoring is the pre-processing stage. This phase involves noise reduction and signal enhancement. Environmental data can be plagued by noise from various sources, including electronic sensor interference, ambient conditions, and human activity. Techniques like filtering, averaging, and Fourier transforms help mitigate these issues. For example, low-pass filters can be employed to remove high-frequency noise from temperature data, while averaging might be used to smooth out fluctuations in pollutant concentration readings.

Following pre-processing, feature extraction is the next crucial step. This process aims to identify and quantify significant patterns or features within the signal. In environmental monitoring, such features might include the amplitude of seismic waves, the frequency of oscillations in atmospheric pressure, or the periodicity of temperature changes. Feature extraction often employs techniques like wavelet transforms, which can decompose a signal into its constituent components, enabling the identification of transient events and trends that are pivotal for environmental assessment.

Once significant features have been extracted, the data is subjected to various analysis and interpretation methodologies. Machine learning and statistical techniques are increasingly employed to make sense of complex environmental data. These approaches can classify, cluster, and predict environmental phenomena based on historical data and detected patterns. For instance, machine learning algorithms can predict weather changes by analyzing trends in atmospheric data or detect anomalies in air quality that suggest the presence of new pollutants.

Time-series analysis is another vital tool in environmental monitoring signal processing. Environmental data is inherently temporal, with variables changing continuously over time. Techniques like Auto-Regressive Integrated Moving Average (ARIMA) models and Seasonal Decomposition of Time Series (STL) allow for the modeling and forecasting of temporal environmental data. These methodologies are crucial for predicting environmental phenomena such as the onset of natural disasters, seasonal variations in pollutant levels, or long-term climate trends.

Another essential aspect is the spatial analysis of environmental signals. Environmental data is often spatially distributed and requires techniques like geostatistics and spatial interpolation to make sense of it. Methods such as Kriging and Inverse Distance Weighting (IDW) allow for the estimation of environmental variables at unmeasured locations based on data from nearby sensors. This spatial perspective is critical for mapping pollutant dispersion, modeling habitat ranges, or assessing the spread of forest fires.

In modern environmental monitoring, the integration of real-time data processing has become increasingly important. The advent of Internet of Things (IoT) devices and networks has enabled the continuous collection of environmental data on an unprecedented scale. Real-time signal processing techniques are employed to provide immediate feedback and alerts. For example, real-time monitoring of seismic signals can trigger early warning systems for earthquakes, while real-time air quality data can inform public health advisories.

Furthermore, the fusion of data from multiple sensors enhances the reliability and comprehensiveness of environmental monitoring. Multisensor fusion techniques combine data from different types of sensors to provide a more holistic view of the environment. This approach can correct for the limitations of individual sensors, leveraging their complementary strengths to provide a more accurate and robust assessment.

For instance, combining satellite imagery with ground-based air quality sensors can offer a detailed picture of both the vertical and horizontal distribution of pollutants.

In conclusion, signal processing is indispensable in environmental monitoring, transforming raw sensor data into actionable insights. Through a combination of noise reduction, feature extraction, temporal and spatial analysis, and real-time processing, signal processing techniques enable the accurate monitoring, assessment, and prediction of environmental phenomena. As environmental challenges continue to escalate, the advancements in signal processing will be crucial in equipping us with the tools needed to understand and address these issues effectively.

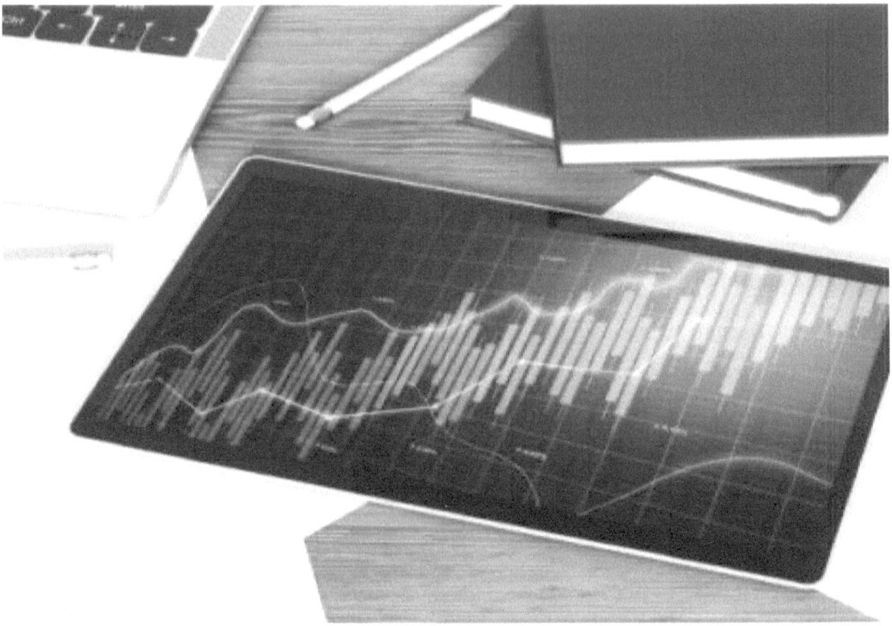

10. Signal Processing and Data Analysis

Signal processing and data analysis are pivotal components in modern technology and scientific research. Whether it's improving the clarity of an audio recording, interpreting data from a distant satellite, or forecasting market trends, these fields form the backbone of numerous applications across various domains. In this chapter, we embark on a detailed exploration of the fundamental principles and advanced techniques that underpin signal processing and data analysis. We will examine both the theoretical underpinnings and practical implementations, providing a comprehensive understanding that caters to both novices and seasoned practitioners.

We start by demystifying the basic concepts and mathematical foundations necessary for grasping the complexities of signal processing. Key topics such as Fourier transforms, filtering, and spectral analysis will be dissected to reveal their roles in transforming raw data into meaningful insights. Then, we delve into the realm of data analysis, highlighting statistical methods, machine learning algorithms, and visualization techniques integral to interpreting and leveraging data effectively.

By the end of this chapter, readers will not only appreciate the intricacies involved in processing signals and analyzing data but also gain the skills required to apply these techniques to real-world problems. This journey through the multifaceted landscape of signal processing and data analysis will equip you with the knowledge and tools to drive innovation and make informed decisions in your respective fields.

10.1 Understanding Data Analysis Techniques and Their Impact on Signal Processing

Signal processing, an indispensable aspect of modern technology, integrates a variety of data analysis techniques to interpret, modify, and utilize signals for diverse applications. In order to comprehend the spectrum of these techniques and how they influence signal processing, it's essential to explore their theoretical foundations, practical methodologies, and real-life applications.

The cornerstone of signal processing lies in its ability to transform signals, which are representations of time-varying or spatially-varying quantities, into valuable information. These signals can originate from myriad sources, encompassing audio, video, radar, and sensor data, to name a few. The efficacy of any signal processing task hinges on the adeptness with which data analysis techniques are applied. Key data analysis techniques that significantly impact signal processing include Fourier Transform, Wavelet Transform, Principal Component Analysis (PCA), and Machine Learning Algorithms.

The Fourier Transform (FT) is perhaps the most seminal technique in signal processing. By converting a signal from its time domain to its frequency domain, the FT enables the identification of periodicities and dominant frequency components within a signal. This technique is crucial for tasks such as filtering, compression, and spectral analysis. In audio processing, for instance, the FT helps in separating different musical notes or in noise reduction. Its impact is seen in how efficiently one can isolate and modify specific frequency components to enhance audio quality or to extract meaningful features from a mixed signal.

The Wavelet Transform (WT) has emerged as a versatile tool, especially in the analysis of non-stationary and transient signals. Unlike FT which uses a single basis function, WT employs a set of basis functions that are scalable and translatable. This property makes WT adept at multi-resolution analysis, where signals are examined at various levels of detail. In image processing,

WT aids in edge detection, compression, and denoising, significantly influencing how images are enhanced and analyzed. Its ability to localize in both time and frequency domains makes it invaluable for medical signal processing, such as in electrocardiograms (ECGs), where it helps isolate sudden changes or irregularities in heartbeats.

Principal Component Analysis (PCA) is another pivotal technique, especially useful in dimensionality reduction and feature extraction. PCA transforms correlated variables into uncorrelated principal components, which can be used to reveal the underlying structure of the data. This technique is instrumental in compressing data while retaining its essential features, making it indispensable in tasks like face recognition in video processing. By reducing dimensionality, PCA not only speeds up the processing time but also mitigates the effects of noise, thereby enhancing the accuracy of signal interpretation.

Machine Learning Algorithms have revolutionized signal processing by introducing adaptive and predictive capabilities. Techniques such as Artificial Neural Networks (ANNs), Support Vector Machines (SVMs), and Convolutional Neural Networks (CNNs) are leveraged to model complex, non-linear relationships in data. In the realm of speech recognition, for instance, ANNs are utilized to map audio signals to phonetic units, thus enabling accurate transcription. The impact of these algorithms is profound, as they improve with experience, progressively increasing the precision and reliability of signal processing tasks.

One illustrative example of the integration of these techniques is in the field of radar signal processing. Radar systems must detect, track, and classify objects within various environments. The Fourier Transform is used to analyze the Doppler shift of signals, distinguishing stationary objects from moving ones. Wavelet Transform assists in resolving radar pulses that may overlap in time but differ in frequency content. PCA can reduce the dimensionality of received signal data, making it easier to classify objects based on features extracted from the signal. Machine Learning Algorithms further refine this classification task by learning patterns from historical data, thus improving object detection and tracking accuracy over time.

In conclusion, the diverse range of data analysis techniques deployed in signal processing each provides unique advantages and improvements.

Whether it's the Fourier Transform's frequency domain analysis, the Wavelet Transform's time-frequency localization, PCA's dimensionality reduction, or the adaptive learning capabilities of Machine Learning Algorithms, these techniques collectively enhance our ability to process and interpret signals in increasingly sophisticated ways. The impact of these techniques is measured not only in improved signal quality and reduced noise but also in the advanced capabilities they afford across various technological domains.

10.2 Adapting Signal Processing Techniques to Big Data

Signal processing techniques are fundamental tools for analyzing and interpreting a variety of data forms, ranging from audio and image signals to complex communications systems. As the volume, variety, and velocity of data have grown in the era of big data, traditional signal processing methodologies have had to evolve. Adapting signal processing techniques to big data involves overcoming several challenges and leveraging new opportunities that arise from massive datasets.

One of the principal challenges in adapting signal processing methods to big data is the sheer size of the data. Traditional algorithms, which were designed for smaller datasets, often become computationally infeasible. For instance, the Fast Fourier Transform (FFT), a widely used technique in signal processing, might struggle with memory and computational bottlenecks when applied to data streams in the terabyte range. Therefore, scalability becomes an essential feature to be considered in the redesign or adaptation of these techniques. Parallel processing frameworks like Apache Hadoop and Apache Spark are often utilized to address these concerns. These frameworks can distribute the computation across a cluster of machines, enabling the processing of large datasets within reasonable timeframes.

Another key consideration is the variety of data that constitutes big data. Signal processing techniques traditionally deal with structured data; however, modern datasets include a broad spectrum of data types, such as text, images, audio, and video. This multimodal nature requires a more flexible approach. For instance, treating text data through NLP (Natural Language Processing) pipelines and time-series data through conventional signal processing techniques in a unified framework is an ongoing research area. Integrating these diverse data types necessitates the enhancement of signal processing algorithms to be versatile enough to handle heterogeneous data sources effectively.

Velocity, or the high rate at which data is generated and must be processed, is another factor compelling the adaptation of signal processing techniques to big data. Real-time or near-real-time processing is increasingly required in applications such as online advertising, stock trading, and network security monitoring. Traditional batch processing models are generally inadequate here. Instead, streaming data platforms like Apache Kafka and frameworks like Apache Flink are becoming more prevalent. These systems are designed to handle continuous flows of data, offering mechanisms to process streams as they arrive. The modification of signal processing algorithms to function in these streaming environments is crucial.

The inherent noise and redundancy present in large datasets further complicate the adaptation process. In conventional settings, denoising techniques like wavelet transforms and Principal Component Analysis (PCA) are employed to filter out irrelevant information. However, with the massive scale of modern datasets, these methods need to be optimized for performance. Techniques such as compressed sensing and random matrix theory have shown promise in efficiently dealing with redundancy and noise in large datasets.

Moreover, machine learning and deep learning have started playing an increasingly important role in the domain of signal processing for big data. Neural networks, particularly convolutional neural networks (CNNs) and recurrent neural networks (RNNs), have been successfully applied to complex signal processing tasks in audio, image, and video data. These techniques can automatically learn to extract features, detect patterns, and make predictions, often outperforming traditional methods. However, the integration of machine learning models into signal processing pipelines introduces new challenges, such as the need for massive labeled datasets for training and the computational overhead of model inference.

Furthermore, privacy and security become critical issues when dealing with big data. Signal processing techniques must now incorporate mechanisms for data anonymization, encryption, and secure multiparty computation to protect sensitive information. These considerations are particularly important in sectors like healthcare and finance, where data breaches can have severe consequences.

In summary, adapting signal processing techniques to big data involves addressing challenges related to scalability, data variety, processing velocity, noise reduction, and privacy. Leveraging distributed computing frameworks, integrating multimodal data handling, adopting real-time processing paradigms, employing advanced mathematical techniques, and incorporating machine learning models are crucial steps in this adaptation process. As data continues to grow in scale and complexity, ongoing research and development in adapting signal processing methodologies will be essential to harness the full potential of big data.

10.3 The Role of Signal Processing in Machine Learning

Signal processing plays a critical role in the field of machine learning, serving as an essential tool for the pre-processing, transformation, and enhancement of data used in various machine learning models. At its core, signal processing involves the analysis, modification, and synthesis of signals such as sound, images, and biological measurements. When integrated with machine learning algorithms, signal processing techniques help improve the accuracy, efficiency, and reliability of predictive models.

One of the primary contributions of signal processing in machine learning is data pre-processing. Raw data collected from sensors or other sources often contain noise and irrelevant information, which can negatively impact the performance of machine learning models. Signal processing techniques, such as filtering, denoising, and normalization, help clean and refine the data. For instance, in audio signal processing, methods like spectral subtraction and Wiener filtering are commonly used to reduce background noise and enhance the quality of the audio signal, making it more suitable for tasks such as speech recognition and speaker identification.

Feature extraction is another area where signal processing significantly contributes to machine learning. Features are specific characteristics or attributes extracted from raw data, which serve as inputs to machine learning models. Signal processing methods enable the extraction of salient features that are crucial for accurate model performance. For example, in image processing, techniques like edge detection, wavelet transforms, and Fourier transforms are employed to extract meaningful features such as edges, textures, and frequency components. In the context of time-series data, which are prevalent in domains like finance and healthcare, signal processing techniques like autoregressive models and Fourier analysis help capture temporal patterns and trends.

Signal processing also aids in dimensionality reduction, a crucial step in managing large datasets. High-dimensional data can pose challenges in terms of computational efficiency and model overfitting. Techniques such as principal component analysis (PCA) and independent component analysis

(ICA) are employed to reduce the dimensionality of data while retaining the most informative components. This not only simplifies the learning process but also enhances the generalization capability of the machine learning models.

Additionally, signal processing enhances the interpretability of machine learning models, particularly in domains requiring transparency and explainability. Techniques like time-frequency analysis and wavelet transforms provide intuitive representations of data, allowing researchers to better understand the underlying patterns and structures. This is particularly important in applications like medical diagnosis, where the interpretability of models can impact clinical decision-making.

In real-time applications, the role of signal processing becomes even more pronounced. Real-time signal processing techniques enable the immediate analysis and modification of incoming data, which is critical for applications such as autonomous driving, real-time video analytics, and online monitoring systems. These techniques ensure that machine learning models receive timely and relevant data, thereby enhancing their responsiveness and accuracy.

Moreover, the integration of signal processing with machine learning facilitates the development of advanced algorithms capable of handling complex and dynamic environments. For instance, in adaptive filtering, signal processing techniques are used to continuously update the model parameters based on the incoming data, allowing the system to adapt to changing conditions. This adaptability is essential for applications like adaptive noise cancellation and adaptive beamforming in communications systems.

Finally, signal processing contributes to the robustness and resilience of machine learning models. By employing techniques like robust filtering and outlier detection, signal processing methods help mitigate the impact of anomalies and disturbances in the data. This is particularly important in ensuring the reliability and stability of machine learning models deployed in critical applications such as financial forecasting, industrial automation, and security systems.

In conclusion, signal processing serves as a foundational element in the machine learning pipeline, enhancing the quality, efficiency, and interpretability of data used in various models. Its techniques are indispensable for data pre-processing, feature extraction, dimensionality reduction, real-time processing, and adaptive learning, making it a vital component in the advancement of machine learning technologies.

10.4 Case Studies of Data-Driven Signal Processing Strategies

In today's era of advanced analytics and big data, the application of data-driven techniques in signal processing has become increasingly critical. The convergence of machine learning, statistical analysis, and signal processing has engendered a host of innovative methodologies. Here we delve into several case studies that illuminate the practical application of these strategies across different domains, illustrating their transformative potential.

Case Study 1: Predictive Maintenance in Industrial Systems

One of the most prominent examples of data-driven signal processing is found in predictive maintenance of industrial machinery. Traditional methods of preventative maintenance relied heavily on scheduled inspections and fixed intervals of parts replacement, often leading to either over-maintenance or unexpected breakdowns. By employing data-driven signal processing strategies, such as vibration analysis and frequency spectrum studies, industries can predict mechanical failures with greater accuracy.

For instance, machine learning algorithms can be trained on historical vibration data from industrial motors. Signal features, extracted through techniques like the Fast Fourier Transform (FFT), serve as inputs to predictive models. These models identify patterns indicative of wear and tear, enabling preemptive interventions. This approach significantly reduces downtime and maintenance costs while enhancing the operational lifespan of machinery.

Case Study 2: Healthcare and Biomedical Signal Processing

In healthcare, the analysis of biomedical signals such as ECG (Electrocardiograms), EEG (Electroencephalograms), and EMG (Electromyograms) has been revolutionized by data-driven approaches. Traditional signal processing methods often fell short in handling the complexity and variability inherent in biomedical signals.

Consider the diagnosis of arrhythmia using ECG signals. Machine learning models, when trained on large datasets of annotated ECG signals, can detect anomalies more precisely than manual interpretation. Feature extraction methods like wavelet transforms help to isolate critical moments within the signal, while classification algorithms such as Support Vector Machines (SVM) and Neural Networks can distinguish between various types of arrhythmias. This results in faster, more reliable diagnostics, enhancing patient care through timely and accurate detection of cardiac issues.

Case Study 3: Environmental Monitoring and Disaster Management

Environmental monitoring systems benefit significantly from data-driven signal processing strategies. For example, seismic signal analysis facilitates the early detection of earthquakes, potentially saving lives through timely alerts. Traditional seismic analysis involved manual interpretation of signal data, but recent advancements have incorporated machine learning to automate this process.

By analyzing seismic waveforms, data-driven models can detect shifts and patterns indicative of an impending earthquake. Techniques such as Principal Component Analysis (PCA) reduce the dimensionality of seismic data, making it more manageable and improving the accuracy of predictions. Additionally, real-time data processing algorithms enable continuous monitoring and faster response times.

Case Study 4: Audio Signal Processing for Speech Recognition

Speech recognition technologies have seen remarkable advancements through data-driven signal processing strategies. Classic methods of speech recognition struggled with background noise and variability in human speech. Modern techniques leverage deep learning algorithms and large-scale datasets to enhance accuracy and reliability.

For instance, the application of Convolutional Neural Networks (CNNs) and Recurrent Neural Networks (RNNs) to audio signal processing has created models capable of understanding and transcribing spoken language in real time. Feature extraction methods such as Mel-Frequency Cepstral Coefficients (MFCC) and spectrogram analysis help in capturing the nuances of human speech, which are then processed by deep learning models to produce accurate text outputs. These advancements are instrumental in the development of virtual assistants, automated transcription services, and improved accessibility for individuals with disabilities.

Case Study 5: Financial Market Analysis

In financial markets, the analysis of time-series data for decision-making has been greatly enhanced by data-driven signal processing strategies. Traditional methods heavily relied on statistical models and human intuition. Today's market analysts utilize machine learning algorithms to dissect and interpret market signals with higher precision.

By processing signals such as stock price movements and trading volumes, models can identify trends and predict future market behaviors. Feature extraction methods like moving averages and technical indicators serve as inputs for predictive models such as Long Short-Term Memory (LSTM) networks. These models can forecast stock prices, aiding traders and investors in making informed decisions.

Case Study 6: Signal Processing in Communications

In the field of communications, data-driven signal processing strategies are employed to optimize signal transmission and reception. For instance, MIMO (Multiple Input Multiple Output) systems in wireless communication use complex algorithms to enhance data throughput and reduce signal interference.

Machine learning techniques are applied to model and predict the behavior of wireless channels, adapt to changing conditions, and optimize power usage. Techniques such as reinforcement learning allow communication systems to dynamically adjust their parameters for optimal performance.

Case Study 7: Smart Grid and Energy Management

Smart grid technology incorporates data-driven signal processing to manage and distribute electricity efficiently. By analyzing signal data from various sensors throughout the grid, algorithms can predict energy demand, detect faults, and optimize the distribution of electricity.

For example, load forecasting models use historical consumption data and real-time sensor data to predict future energy usage. This predictive capability enables utility companies to manage resources more effectively and respond proactively to potential overloads or failures, ensuring a stable and efficient energy supply.

These case studies demonstrate the diverse applications of data-driven signal processing strategies across various sectors. They highlight the potential of these methodologies to enhance operational efficiency, improve predictive capabilities, and ultimately drive innovation. By integrating advanced signal processing techniques with machine learning and statistical analysis, industries can unlock new levels of performance and reliability.

#1
#2

11. Structuring Teams for Signal Processing Success

Achieving excellence in signal processing is not merely a function of superior algorithms and cutting-edge technology. It is equally contingent on how teams are structured and how effectively they can collaborate towards common goals. In this chapter, we delve into the intricacies of assembling, organizing, and empowering teams to maximize their potential in the demanding field of signal processing. By examining case studies, best practices, and theoretical frameworks, we will uncover the strategies that pave the way for success. Whether you are a project manager aiming to optimize team performance or an individual contributor seeking to understand your role within a larger system, this chapter offers valuable insights tailored to enhance both collective and individual achievements in signal processing projects.

11.1 Organizing Teams for Effective Signal Processing Management

Effective signal processing management is fundamentally dependent upon the organization of teams. Organizing teams for this task requires a structured approach that emphasizes clearly defined roles, specialized expertise, and collaborative synergy. Ensuring that these elements are harmoniously integrated significantly boosts operational efficiency and the quality of outcomes.

First and foremost, delineating roles and responsibilities is crucial. In a well-organized signal processing team, there are typically several distinctive roles including project managers, signal processing engineers, data analysts, software developers, and quality assurance specialists. Each role is essential and contributes uniquely to the overarching goal. Project managers are responsible for overseeing the project timeline, setting milestones, and ensuring that all team members meet their objectives within stipulated deadlines. They also handle resource allocation, stakeholder communication, and risk management.

Signal processing engineers, on the other hand, carry the core technological responsibilities. Their expertise lies in designing, implementing, and optimizing algorithms that process signals to extract meaningful data or transform it into a desired state. They must understand both theoretical principles and practical applications, ensuring that the algorithms they work on are both innovative and reliable.

Data analysts play a critical role in interpreting the processed signals. They delve into the data post-processing to identify patterns, generate insights, and create visualizations that communicate complex information in an accessible manner. Their analyses guide strategic decisions and future project directions, making their role indispensable.

Software developers contribute to the infrastructure that supports the signal processing operations. They develop and maintain the software tools and platforms that enable the signal processing algorithms to function seamlessly. Their work ensures that the system is not only robust but also scalable and user-friendly.

Quality assurance specialists are tasked with maintaining high standards across all processes. They rigorously test algorithms, software, and data outputs to identify any discrepancies, bugs, or deviations from expected outcomes. Through their meticulous work, they uphold the reliability and credibility of the processing system.

In addition to clearly defined roles, expertise within these roles must be specialized. Signal processing is a multifaceted domain that integrates aspects of mathematics, computer science, and engineering. Therefore, each member of the team must possess deep knowledge and proficiency in their respective areas. Regular training and professional development opportunities are essential to keep team members abreast of the latest advancements and best practices in the field.

Embedding a culture of collaboration is another cornerstone for organizing effective teams. While each team member has specialized expertise, the collective success of the project hinges on their ability to collaborate seamlessly. This requires fostering an environment where open communication is encouraged, and interdisciplinary knowledge is shared. Regular team meetings, brainstorming sessions, and collaborative workshops can facilitate this exchange, ensuring that all members are aligned with the project's objectives and methodologies.

In addition to fostering collaboration, leveraging cutting-edge tools and technologies can elevate the team's performance. Adopting advanced software for project management, communication platforms, and data processing can streamline workflows and enhance productivity. Tools such as MATLAB, Python, and specialized signal processing libraries can significantly reduce the time and effort needed to develop and test algorithms.

Effective signal processing management also requires a strategic approach to problem-solving. Teams must be equipped with the methodologies to tackle challenges systematically. Employing frameworks such as Agile or Six Sigma can bring structure to the problem-solving process, ensuring that issues are addressed promptly and efficiently.

Moreover, continuous evaluation and feedback mechanisms must be embedded within the team's operations. Establishing key performance indicators (KPIs) and regular review cycles help monitor progress and identify areas for improvement. Feedback loops, both internal and external, are vital for continuous growth and adaptation.

Intrinsic to organizing teams for effective signal processing management is the leadership's vision and direction. Leaders must not only possess technical acumen but also the ability to inspire and motivate the team. They are responsible for fostering a positive culture, driving innovation, and ensuring alignment with the organization's goals.

In conclusion, organizing teams for effective signal processing management involves a multi-dimensional approach focusing on defined roles, specialized expertise, collaborative culture, use of advanced tools, strategic problem-solving, continuous evaluation, and strong leadership. When these elements coalesce, they create a dynamic and efficient team capable of achieving exceptional outcomes in signal processing projects.

11.2 Centralization vs. Decentralization in Signal Processing

In signal processing, the debate between centralization and decentralization is pivotal to understanding the design and efficiency of various systems, such as those used in telecommunications, data transmission, and sensor networks. Each approach comes with its distinct advantages and challenges, shaped by the specific requirements of the application at hand.

Centralization in signal processing means that the processing of signals is conducted at a single centralized point. This centralized point typically involves a powerful processor or a cluster of processors in a robust computing facility. In centralized systems, raw data from multiple sources is gathered and transmitted to a central processor for analysis and decision-making. The primary advantage of centralization is the ability to leverage significant computational power, which can be utilized for complex algorithms that might be infeasible in distributed systems.

Moreover, centralized systems allow for easier maintenance and upgrades since all significant operations occur at a singular location. Centralization also permits a more coherent and uniform decision-making process due to the holistic view of the entire network or system. For instance, in radar signal processing, centralized systems can aggregate data from numerous sensors, providing comprehensive analysis and enhanced detection capabilities. Also, centralized architectures often benefit from economies of scale, in which the cost per unit of processing power decreases as more resources are pooled together.

However, there are notable drawbacks to centralization. The main issue is that such systems can become a single point of failure. If the central processing unit or facility encounters a problem, the whole system could be rendered inoperative. Scalability is another concern; as the system grows and more data inputs are added, the central processor might become overwhelmed, leading to delays and reduced performance. Additionally, centralized systems can face significant latency issues because the raw data must often travel long distances to reach the processing center. This can be particularly problematic in real-time applications where delay is critical.

On the other end of the spectrum, decentralization in signal processing involves distributing the processing tasks across multiple nodes or processors spread out within the system. Each node usually processes its own local data, making decisions independently based on this localized information. This architecture is notably used in distributed sensor networks, peer-to-peer systems, and certain telecommunication systems.

One of the primary benefits of decentralization is its inherent robustness. Since processing is distributed, the failure of one node does not incapacitate the entire system. Decentralized systems are also typically more scalable since new nodes can be added without significantly increasing the load on existing nodes. They offer the advantage of reduced latency as local data is processed locally, minimizing the need for long-distance data transmission.

Additionally, decentralized systems tend to be adaptable and resilient to changes in the environment or system configuration. For example, in a military communication network, the ability to function despite the failure of individual nodes is critical. Similarly, decentralized processing in sensor networks can lead to energy savings since only relevant data is processed and transmitted, conserving power and bandwidth.

On the downside, decentralization presents challenges in terms of coordination and data consistency. Since individual nodes operate semi-independently, ensuring that all parts of the system work harmoniously can be difficult. There is also the issue of redundancy; multiple nodes might process the same data, leading to inefficient use of resources. In some scenarios, complex algorithms required for certain signal processing tasks might be impractical to implement on smaller, less powerful processors typical of decentralized systems. Furthermore, ensuring security and privacy can be challenging, as more points of access might be exploited maliciously.

In conclusion, both centralization and decentralization present unique advantages and inherent challenges in the realm of signal processing. The choice between the two approaches depends on the specific requirements of the application, such as the need for robustness, scalability, latency sensitivity, and the complexity of the signal processing tasks. Carefully balancing these factors is key to designing a successful signal processing system.

11.3 Managing Cross-Functional Collaboration in Signal Processing

Effective cross-functional collaboration in signal processing is imperative for the successful realization of complex projects that often require the integration of diverse skills and perspectives. Signal processing is a field that spans numerous applications, including telecommunications, audio engineering, medical imaging, and more, each requiring specialized knowledge. Hence, managing collaboration among professionals from different disciplines is essential for innovation and problem-solving.

Central to cross-functional collaboration is the clear definition of roles and responsibilities. Project managers must delineate who is responsible for what, ensuring that each team member understands their specific duties and how these fit into the broader objectives. A clear organizational structure helps prevent overlap of efforts and minimizes the risk of important tasks being neglected.

Communication is another cornerstone of effective collaboration. Regular meetings, both formal and informal, should be held to discuss progress, challenges, and upcoming tasks. Tools like Slack, Microsoft Teams, or Asana can facilitate better communication by providing centralized platforms for sharing updates, documents, and quick queries. Additionally, establishing a common vocabulary around signal processing concepts and terms can significantly reduce misunderstandings. This is particularly important when team members come from diverse backgrounds, such as electrical engineering, computer science, and applied mathematics.

Interdisciplinary training and knowledge sharing can also enhance collaboration. Offering workshops, seminars, or even informal lunch-and-learn sessions on fundamental concepts of signal processing can help bridge knowledge gaps. Encouraging team members to share their expertise through presentations or documentation can foster a culture of mutual learning. This not only improves the team's overall competency but also strengthens the collaborative spirit.

Project documentation is another critical element. Detailed, well-organized documentation of project plans, milestones, and individual contributions can streamline cross-functional collaboration. It provides a single source of truth, ensures continuity when team members rotate or leave, and helps new members get up to speed quickly. Version control systems like Git can help manage changes to project documentation, source code, and algorithms, ensuring that updates are tracked and that previous versions can be referenced if needed.

Integrating tools that support various aspects of signal processing can also enhance cross-functional collaboration. Software like MATLAB or Python's SciPy library can be employed for simulations and algorithm development. Meanwhile, project management tools like JIRA or Trello can help track tasks and milestones. Ensuring that all team members are proficient with these tools can streamline collaboration significantly.

Conflicts are an inevitable part of any collaborative effort, especially in cross-functional teams where differing viewpoints and approaches are common. Addressing conflicts promptly and constructively is essential. Conflict resolution training for team leaders can be beneficial. Techniques such as active listening, mediation, and seeking compromise should be employed to ensure that disagreements do not derail the project.

Leadership plays a pivotal role in managing cross-functional teams in signal processing. Leaders must not only possess a sound understanding of signal processing but also exhibit strong interpersonal and project management skills. They should set a collaborative tone from the outset and lead by example, demonstrating respect, open-mindedness, and a willingness to listen to different perspectives.

Innovation thrives in a collaborative environment where different perspectives are valued. In signal processing, where advances in algorithm efficiency, hardware implementation, and application-specific adaptations are constantly pursued, fostering an innovative environment is paramount. Encouraging a culture where team members feel comfortable proposing unconventional ideas or questioning established methods can lead to significant breakthroughs.

Moreover, recognizing and rewarding collaborative efforts is crucial. Incentives should not only be based on individual contributions but also on how well team members work together to achieve project goals. Public recognition, team awards, and professional development opportunities can motivate team members to invest in collaboration.

In summary, managing cross-functional collaboration in signal processing involves a multifaceted approach. Defining clear roles, fostering effective communication, promoting interdisciplinary training, maintaining comprehensive documentation, integrating appropriate tools, adeptly resolving conflicts, exercising strong leadership, and recognizing collaborative efforts are all critical factors. When these elements are successfully implemented, they create a cohesive, innovative environment where multidisciplinary teams can thrive and achieve remarkable advancements in signal processing technology.

11.4 Governance and Compliance in Signal Processing Practices

In the evolving landscape of technology, governance and compliance have become crucial elements of signal processing practices. Signal processing, the science of analyzing, modifying, and synthesizing signals such as sound, images, and biological measurements, is central to various applications that influence numerous sectors, including telecommunications, healthcare, and automotive industries. With the rapid advancements in this field, ensuring robust governance and compliance frameworks is critical to address legal, ethical, and social implications.

Governance in signal processing involves establishing policies, procedures, and standards to guide the decision-making processes and operational practices within an organization. It ensures that signal processing activities align with the organization's objectives, regulatory requirements, and industry best practices. Good governance practices help mitigate risks, enhance transparency, and foster accountability among stakeholders. For instance, an organization involved in developing signal processing algorithms for medical devices must adhere to stringent guidelines set by regulatory bodies such as the U.S. Food and Drug Administration (FDA) to ensure the safety and efficacy of their products.

Compliance, on the other hand, refers to adhering to laws, regulations, standards, and ethical guidelines applicable to signal processing activities. Compliance ensures that signal processing practices do not infringe on legal rights and conform to the ethical standards of the profession. Non-compliance can lead to legal repercussions, financial penalties, and reputational damage. In the European Union, the General Data Protection Regulation (GDPR) imposes strict requirements on how personal data is processed and stored, directly impacting signal processing techniques that handle such data..

A key aspect of governance in signal processing is data governance, which encompasses the management of data availability, usability, integrity, and security. Effective data governance frameworks ensure that data used in signal processing is accurate, consistent, and protected against unauthorized access. This is particularly important in applications like speech recognition and biometric systems, where data quality and security are paramount. Organizations must implement data governance policies that define roles and responsibilities, establish data standards, and enforce data protection measures to safeguard sensitive information.

Moreover, ethical considerations play a significant role in governing signal processing practices. Ethical guidelines help address issues related to privacy, bias, and the potential misuse of signal processing technologies. For instance, in the field of surveillance, signal processing technologies such as facial recognition systems raise critical ethical concerns regarding privacy and surveillance. Organizations must ensure that their technologies comply with ethical standards to prevent intrusive monitoring and discriminatory practices. Adopting ethical guidelines and conducting regular ethical audits can help in identifying and mitigating such risks.

Risk management is another crucial component of governance in signal processing. Organizations must identify, assess, and manage risks associated with their signal processing activities. This involves implementing robust risk management frameworks that can address various types of risks, including operational, legal, and reputational risks. For example, in autonomous driving systems, signal processing algorithms play a vital role in interpreting sensor data to make real-time decisions. Any failure in these algorithms can lead to catastrophic consequences, highlighting the importance of rigorous risk management.

In compliance, standardization is critical to ensure uniformity and interoperability in signal processing practices. Standards developed by professional bodies such as the Institute of Electrical and Electronics Engineers (IEEE) provide guidelines for the development and implementation of signal processing systems. Adhering to these standards ensures that signal processing technologies are reliable, interoperable, and maintainable. Compliance with established standards also facilitates collaboration and innovation within the industry, driving technological advancements.

Training and awareness are pivotal to fostering a culture of compliance within organizations. Employees involved in signal processing must be educated about the relevant laws, regulations, and ethical standards. Regular training programs and workshops can help in keeping the workforce informed about the latest compliance requirements and best practices. Additionally, establishing a compliance monitoring system can help organizations track adherence to compliance policies and identify areas needing improvement.

In conclusion, governance and compliance are integral to the responsible practice of signal processing. By implementing robust governance frameworks and adhering to compliance requirements, organizations can mitigate risks, ensure ethical conduct, and maintain the integrity and reliability of their signal processing activities. As signal processing continues to advance, the importance of governance and compliance will only grow, necessitating continuous efforts to uphold these principles.

12. Sustainability in Signal Processing

In recent years, sustainability has become a critical focus across multiple disciplines, driven by the pressing need to address environmental challenges and promote responsible resource management. Signal processing, a field that underpins numerous technologies ranging from telecommunications to medical imaging, is no exception to this imperative. This chapter delves into the evolving landscape of sustainability within signal processing, exploring innovative methodologies and practices designed to minimize ecological footprints while maintaining high standards of performance and efficiency.

We begin by examining the fundamental principles of green engineering as they apply to signal processing. Key considerations such as energy efficiency, resource optimization, and lifespan extension of electronic components are scrutinized to understand their impacts and benefits. The discussion progresses to contemporary advancements in low-power design techniques, adaptive algorithms, and resource-aware signal processing frameworks that significantly reduce energy consumption.

Through case studies and real-world applications, the chapter illuminates the transformative efforts of researchers and practitioners committed to integrating sustainability into their work. These examples not only showcase the tangible benefits of sustainable signal processing but also highlight the challenges and trade-offs encountered in this endeavor. By presenting a comprehensive overview of current trends, technological innovations, and future directions, this chapter aims to equip professionals and academics with the insights needed to contribute to a more sustainable technological ecosystem.

Ultimately, the pursuit of sustainability in signal processing is not merely an option but a necessity, aligned with global efforts to combat climate change and conserve finite resources. This chapter underscores the importance of interdisciplinary collaboration and continuous innovation in achieving a balance between technological advancement and ecological stewardship.

12.1 Integrating Sustainability into Signal Processing Strategies

Integrating sustainability into signal processing strategies represents a forward-thinking approach that not only addresses the technical requirements of modern systems but also responds to the growing imperative for eco-friendly practices. Signal processing, being integral to various domains such as telecommunications, medical imaging, audio engineering, and more, consumes significant computational resources and energy. Consequently, it is crucial to consider sustainable methodologies within this field to reduce its environmental footprint.

One practical step toward sustainable signal processing involves optimizing algorithm efficiency. Algorithms are the backbone of signal processing tasks, and their complexity directly influences energy consumption. Developing and implementing algorithms that require fewer computational cycles can substantially cut down energy usage. For instance, signal compression techniques that reduce the amount of data needing processing without compromising quality can lead to lower energy consumption and extended hardware longevity. By focusing on algorithmic simplicity and efficiency, engineers can make substantial strides toward sustainability.

Energy-efficient hardware design is another critical component of sustainable signal processing strategies. As the execution of signal processing tasks often requires considerable computational power, it is essential to choose hardware components designed for minimal energy consumption. Advances in semiconductor technology, such as low-power transistors and energy-efficient processors, can play a significant role in this realm. Additionally, employing specialized hardware accelerators like Field-Programmable Gate Arrays (FPGAs) and Graphics Processing Units (GPUs), which can execute parallel computations more efficiently than general-purpose CPUs, can further enhance energy savings.

Implementing adaptive and dynamic resource management techniques also contributes to sustainability. Dynamic Voltage and Frequency Scaling (DVFS) is a widely researched method that adjusts the processor's voltage and frequency according to workload demands. By scaling down during low-demand periods, the system conserves energy, whereas it ramps up performance when necessary. This dynamic adjustment ensures that energy is not wasted during idle times or under light loads, aligning processing power with actual need.

Moreover, the integration of renewable energy sources to power signal processing infrastructures can profoundly impact sustainability goals. Utilizing solar, wind, or other renewable energies to offset the reliance on fossil fuels not only reduces carbon emissions but also supports the scalability of greener technological ecosystems. Data centers and other facilities that house signal processing operations can be partially or fully powered by renewable energy, which aligns with global sustainability objectives.

The aspect of recyclability and reusability of hardware components should not be overlooked. Signal processing systems often undergo upgrades, leading to the disposal of older equipment. Encouraging the design of modular, recyclable, and upgradable hardware allows for the retention and reuse of equipment, thereby minimizing electronic waste. This circular economy approach not only lowers material consumption but also reduces the energy and resources expended in the production of new hardware.

Software tools used in signal processing can also be optimized for sustainability. Lightweight software with minimalistic coding reduces the need for extensive computational capacity. Furthermore, cloud-based signal processing offers a sustainable alternative by sharing computational resources across multiple users, thus achieving higher resource utilization and reducing overall energy consumption when compared to localized, individual systems.

Collectively, investing in education and research to foster innovations in green signal processing is paramount. Academia and industry collaboration can propel the development of new techniques and tools that prioritize sustainability. Training the next generation of engineers in sustainable practices ensures the perpetuation of eco-friendly approaches within the signal processing field.

In summary, integrating sustainability into signal processing strategies involves a multifaceted approach that includes optimizing algorithms, choosing energy-efficient hardware, employing dynamic resource management, leveraging renewable energy, ensuring hardware recyclability, and developing lightweight software solutions. These combined efforts contribute substantially to reducing the environmental impact of signal processing operations, paving the way for a more sustainable technological future.

12.2 The Role of Ethical Signal Processing in Building Trust

Ethical signal processing has become a cornerstone in developing trustworthy communication and data handling systems in an increasingly digital world. With the proliferation of artificial intelligence (AI) and machine learning (ML), signal processing has taken on new dimensions, encompassing not only the optimization of signals for clarity and accuracy but also ensuring that the processes involved are ethical and transparent.

Signal processing essentially involves the analysis, manipulation, and synthesis of signals—whether these are audio, video, or other forms of data—to extract meaningful information or enhance quality. In the context of ethical signal processing, the focus shifts from mere technical efficiency to the broader implications of these processes on privacy, security, and societal trust.

The role of ethical signal processing in building trust can be bifurcated into three primary aspects: transparency, fairness, and accountability.

Transparency in signal processing instills confidence by making the processes understandable and open to scrutiny. For instance, in speech recognition systems, providing user-friendly documentation and clear descriptions of how data is collected, processed, and utilized can alleviate concerns about unauthorized surveillance or misuse of personal information. Transparency helps users to comprehend the workings of the systems they rely on, fostering a sense of trust through openness.

Moreover, transparency also involves the clear communication of the limitations and capabilities of signal processing systems. This means that users should be made aware of the potential errors, biases, and uncertainties inherent in these systems. By acknowledging imperfections and actively working towards minimizing them, developers can cultivate a more honest relationship with users.

Fairness in signal processing involves ensuring that the algorithms and models employed do not reinforce or create biases. This is particularly crucial in systems that impact significant sectors of society, such as healthcare, criminal justice, and finance. For example, if a facial recognition system demonstrates a higher error rate for people of a certain ethnicity, this can lead to unfair treatment and erode trust in the technology. Addressing these issues requires a concerted effort to audit and refine algorithms continuously, incorporating diverse data sets and perspectives to ensure broad applicability and equitable treatment.

Building fairness into signal processing also involves being proactive about diversity in training data and model validation processes. By doing so, system developers and users can be more confident that the outputs generated are representative and just. This reduces the risk of marginalizing any group and ensures fair access to the benefits of technology.

Accountability is the cornerstone of ethical signal processing, encompassing the responsibility of developers, organizations, and regulators to uphold ethical standards. This means establishing robust mechanisms for detecting and rectifying unethical practices, such as improper data usage or manipulative algorithmic behavior. Accountability also involves maintaining clear channels for users to report issues and grievances, and ensuring that these are addressed promptly and effectively.

Regulatory frameworks play a significant role here, instituting checks and balances through laws and guidelines that enforce ethical practices. Companies need to adopt internal policies aligned with these regulations, emphasizing the importance of ethically sound signal processing in their operations. Auditing processes, both internal and external, can help ensure adherence to ethical standards, providing transparency and fostering trust.

Furthermore, the aspect of accountability extends to the continuous education and training of stakeholders involved in signal processing. By fostering a culture of ethical awareness, organizations can stay ahead of potential ethical dilemmas, encouraging innovation that prioritizes societal well-being alongside technological advancement.

In conclusion, the integration of transparency, fairness, and accountability into signal processing is pivotal in building and maintaining trust among users. As we advance further into an era dominated by digital communication and data processing, these ethical principles will not only serve as the foundation for building trust but also ensure the sustainable and equitable growth of technology. Signal processing systems that prioritize ethical considerations will ultimately lead to more robust, reliable, and trusted technologies, fostering progress that benefits all aspects of society.

12.3 Building Sustainable Signal Processing Models

Sustainable signal processing models have gained prominence in the field of engineering and technology, as the world grapples with the implications of climate change, environmental degradation, and resource depletion. Building these models involves adopting practices and principles that ensure the longevity, efficiency, and minimal environmental impact of signal processing technologies. This endeavor is multidimensional, encompassing aspects ranging from energy consumption and material usage to operational efficiency and lifecycle management.

One of the foremost considerations in developing sustainable signal processing models is energy efficiency. Signal processing operations, particularly in real-time applications, often require significant computational power, which translates to high energy consumption. This energy demand has both economic and environmental repercussions. Minimizing energy consumption can be achieved by optimizing algorithms to be more efficient, reducing the number of arithmetic operations, and utilizing low-power hardware components. Moreover, advanced techniques like adaptive filtering, efficient coding, and compressed sensing can provide high processing performance while conserving energy.

Another critical component is the design of hardware capable of supporting sustainable signal processing. Modern signal processing often relies heavily on specialized hardware such as digital signal processors (DSPs), field-programmable gate arrays (FPGAs), and application-specific integrated circuits (ASICs). These components must be designed with a focus on reducing energy consumption, using sustainable materials, and ensuring long lifecycles. Implementations using nanotechnology and other emerging materials have the potential to further reduce the environmental impact by requiring fewer resources and yielding less electronic waste.

Software optimization plays a vital role in sustainability. Writing efficient code that minimizes computational overhead can substantially reduce power consumption. The adoption of parallel processing and distributed computing can optimize resource utilization. Techniques like dynamic voltage and frequency scaling (DVFS) allow software to adjust the power and performance characteristics of a processor in real-time, depending on the workload, hence conserving energy during lower activity periods.

Furthermore, embracing a lifecycle perspective is essential for sustainability in signal processing models. This involves considering the environmental impact of a system from its inception through its disposal. Practices like modular design and the use of recyclable materials can facilitate easier upgrades and end-of-life disposal, reducing electronic waste. Developing models that can be easily reprogrammed or repurposed for various tasks can extend their useful life and decrease the need for new hardware.

Data management is another dimension where sustainability can be enhanced. Efficient signal processing often generates and relies on vast amounts of data. Proper data management strategies, such as data compression and intelligent storage solutions, can minimize the energy and resources required for data storage and transmission. Cloud computing offers a path to maximize resource utilization through shared infrastructure, but it must be employed judiciously to avoid offsetting these gains through increased cooling and maintenance demands of large data centers.

Additionally, adopting green engineering principles can further the cause of sustainability. This involves designing systems that consider environmental implications right from the start. For instance, selective use of energy-efficient components, recyclable materials, and manufacturing processes with lower environmental footprints can considerably cut down the adverse impacts. Green engineering also emphasizes the importance of maintaining system efficacy over prolonged periods, reducing the need for frequent replacements.

In tandem with technological advancements, the role of regulations and standards cannot be downplayed. Developing and adhering to industry standards that prioritize sustainability can drive systemic changes across the field of signal processing. Standards like the ENERGY STAR for electronics and LEED certification for environmentally sustainable buildings are examples from other industries that can be adapted to fit signal processing models.

Finally, a shift in mindset among engineers, designers, and stakeholders is vital for the sustainability of signal processing systems. It necessitates fostering a culture that values long-term ecological health over short-term gains. Sustainable innovations often require upfront investment and may not yield immediate economic benefits, but their long-term impact on resource conservation and environmental protection is immeasurable.

Building sustainable signal processing models is no longer a theoretical exercise but a pressing practical need. It requires a holistic approach that integrates energy-efficient algorithms, resource-conscious hardware design, lifecycle management, and robust standards. Only through concerted efforts and a commitment to sustainability can the field of signal processing contribute to a more sustainable future.

12.4 Challenges in Balancing Signal Processing and Sustainability Goals

The necessity of balancing signal processing and sustainability goals presents a unique set of challenges that span across multiple domains, from technical constraints to environmental and social considerations. This balance is critical, as signal processing technologies play a crucial role in modern communications, healthcare, and industrial applications, while concurrently, the push for sustainable practices calls for responsible resource utilization and environmental stewardship.

Firstly, one of the primary challenges is the intrinsic energy consumption associated with signal processing systems. Signal processing algorithms and hardware often require significant computational power, which translates to substantial energy usage. For instance, data centers, which are integral to the functioning of modern digital communication systems, consume vast amounts of electricity. This energy demand contributes markedly to carbon emissions, given that a large portion of global electricity generation still relies on fossil fuels. Reducing this energy consumption without compromising the performance and reliability of signal processing tasks is a complex problem that demands innovative solutions, such as the development of more efficient algorithms and low-power hardware designs.

Moreover, the manufacturing processes of signal processing hardware and components pose considerable environmental challenges. The production of semiconductors, for example, involves the use of toxic chemicals and substantial quantities of water, which can lead to pollution and resource depletion if not managed correctly. Addressing these issues requires advancements in green manufacturing practices, including the adoption of less harmful materials, recycling and reusing electronic components, and improving the overall efficiency of production processes.

Beyond the environmental impact, there is also the consideration of electronic waste (e-waste). The rapid pace of technological advancement in signal processing leads to shorter lifespans for devices, thus contributing to the growing problem of e-waste. E-waste contains hazardous materials such as lead, mercury, and cadmium, which can have severe repercussions on human health and the environment if not properly disposed of. Implementing sustainable strategies for e-waste management, such as designing for longevity, improving recyclability, and establishing robust recycling programs, is vital to mitigating these adverse effects.

Economic factors further complicate the balance between signal processing and sustainability. The push for greener technologies often involves higher initial costs, whether it be for research and development or for the implementation of sustainable practices. This can pose a barrier, particularly for smaller companies or those operating in highly competitive markets where cost efficiency is paramount. Nonetheless, integrating sustainable practices can ultimately lead to long-term financial benefits through energy savings, regulatory compliance, and enhanced corporate reputation. Bridging the gap between short-term costs and long-term gains requires a thorough understanding of the economic landscape and a commitment to responsible investment.

On the social front, there is an increasing demand for technological equity and access. Signal processing technologies are pivotal in bridging digital divides and enhancing communication and healthcare services in underserved regions. However, sustainably scaling these technologies to reach marginalized populations without exacerbating environmental impact remains a formidable challenge. This necessitates a multi-faceted approach that includes not only the development of low-power, cost-effective solutions but also the establishment of supportive infrastructures and policies that promote equitable access to technology.

Furthermore, the challenge extends to the realm of education and workforce development. The transition to incorporating sustainable practices in signal processing requires a workforce that is knowledgeable about both advanced signal processing techniques and sustainable engineering principles. This underscores the need for interdisciplinary educational programs and continuous professional development opportunities that equip engineers, scientists, and technologists with the skills needed to innovate responsibly.

In conclusion, the intricate relationship between signal processing and sustainability goals encapsulates a range of challenges that must be carefully navigated. From optimizing energy efficiency and minimizing environmental impact to addressing economic and social considerations, a holistic and forward-looking approach is essential. By fostering innovation in green technologies, promoting sustainable manufacturing and disposal practices, and ensuring equitable access and education, it is possible to achieve a balanced integration of signal processing advancements and sustainability imperatives.

25700
35890
12440
35890
97633
34980
SHUTTLE PASSAGE

13. Signal Processing and Its Societal Impact

In the sprawling landscape of modern technology, few fields have reshaped our daily lives as profoundly as signal processing. From the seemingly mundane actions of making a phone call or streaming a movie, to the life-saving operations in medical imaging and disaster response systems, the fingerprints of signal processing are ubiquitous. As we delve into this chapter, we will explore the intricate mechanisms behind signal processing, unravel its historical evolution, and examine its far-reaching implications on society. We will witness how the transformation of raw data into meaningful information not only powers devices and networks, but also catalyzes advancements that redefine industries and set new paradigms for future innovation. By understanding the principles and applications of signal processing, we can better appreciate its monumental role in shaping the world we live in today and the potential it holds for tomorrow.

13.1 The Societal Responsibility of Signal Processing

The field of signal processing, which involves the analysis, manipulation, and transformation of signals to enhance or extract vital information, has grown to be a cornerstone of many modern technologies. Its applications span across various industries, including telecommunications, healthcare, finance, and entertainment. As signal processing becomes increasingly embedded in our daily lives, it is important to recognize and address the societal responsibilities that come with its advancements.

One critical aspect of societal responsibility in signal processing involves ensuring equitable access to technology. The benefits of advancements in signal processing technologies, such as improvements in medical diagnostics through imaging technologies or enhanced communications through efficient data transmission, should not be limited to privileged groups. Inequitable distribution of these technologies can contribute to a widening digital divide, exacerbating existing socio-economic disparities. Signal processing professionals must work towards developing cost-effective and accessible solutions to ensure that all segments of the population can benefit from these innovations.

Furthermore, the ethical use of signal processing technologies is paramount. As these technologies often deal with sensitive information, such as personal health data or financial records, there is a pressing need to ensure privacy and data security. Ethical considerations must be integrated into the design, development, and deployment phases of signal processing systems. For instance, robust encryption methods should be employed to safeguard data, and data anonymization techniques should be utilized to protect individuals' privacy while still allowing valuable insights to be gleaned from large datasets.

The use of signal processing in surveillance also raises significant ethical concerns. While surveillance technologies can enhance security and aid in crime prevention, they can also infringe upon individuals' privacy and civil liberties if not regulated appropriately. Signal processing professionals must be vigilant about the potential for misuse of these technologies and advocate for policies and frameworks that balance security needs with the protection of individual rights. Transparency in the deployment and usage of surveillance technologies is crucial to maintaining public trust.

In addition to privacy concerns, the environmental impact of signal processing technologies cannot be overlooked. The development and operation of these technologies often require significant computational resources, leading to high energy consumption and a substantial carbon footprint. Signal processing engineers have a responsibility to design algorithms and systems that are not only efficient in terms of performance but also in energy usage. Innovation in low-power signal processing techniques and the development of sustainable computing infrastructures can help mitigate these environmental challenges.

The societal responsibility of signal processing extends to the ethical implications of artificial intelligence (AI) and machine learning (ML), which often rely heavily on signal processing techniques. AI and ML systems are increasingly being used for decision-making in critical areas such as healthcare, law enforcement, and employment. Ensuring that these systems are fair, unbiased, and transparent is crucial to prevent discrimination and uphold justice. Signal processing professionals must engage in ongoing dialogue about the ethical considerations of AI and ML, actively working towards the development of fair algorithms and advocating for the responsible use of technology.

Lastly, education and public awareness play a vital role in fulfilling the societal responsibility of signal processing. Professionals in the field should endeavor to educate the public about the benefits and potential risks associated with signal processing technologies. This can be achieved through public lectures, workshops, and collaborations with educational institutions. By demystifying the technology and fostering an informed public, the societal impact of signal processing can be better understood and managed.

In conclusion, the societal responsibility of signal processing is multifaceted, encompassing equitable access, ethical use, privacy and security concerns, environmental impact, and the ethical implications of AI and ML. As the field continues to evolve and integrate into various aspects of modern life, signal processing professionals must remain committed to addressing these societal challenges. By doing so, they can ensure that the advancements in signal processing contribute positively to society, fostering an inclusive, ethical, and sustainable technological landscape.

13.2 Addressing Social Issues Through Signal Processing Initiatives

Addressing social issues through advanced signal processing initiatives is an approach that leverages the power of technology to tackle a variety of societal challenges. Signal processing, traditionally a sub-discipline of electrical engineering, involves the analysis, modification, and synthesis of signals to improve their quality, extract information, or enable new capabilities. This broad field, which intersects with areas such as machine learning, artificial intelligence, and data science, provides significant opportunities to address pervasive social problems.

At its core, signal processing deals with the transformation of raw data into meaningful insights. This data can come from various sources, including audio, video, medical imaging, and sensor networks. By applying methods to clean, compress, enhance, and interpret these signals, engineers and scientists can create solutions that positively impact areas like healthcare, environmental monitoring, public safety, and social connectivity.

In the realm of healthcare, one notable application of signal processing is in medical imaging and diagnostics. Techniques such as Magnetic Resonance Imaging (MRI), Computed Tomography (CT) scans, and Electrocardiograms (ECGs) rely heavily on signal processing algorithms to render detailed images and analyze physiological signals. Advanced filtering and pattern recognition techniques can detect anomalies and aid in early diagnosis of conditions like cancer, cardiovascular diseases, and neurological disorders. Furthermore, wearable health monitors employ signal processing algorithms to continuously track vital signs, enabling remote patient monitoring and timely medical interventions, thus improving healthcare accessibility and reducing hospital readmissions.

Environmental monitoring is another critical area where signal processing plays a transformative role. Sensors deployed in various environments, from urban areas to remote ecosystems, collect vast amounts of data on parameters such as air quality, water quality, and seismic activity. Signal processing techniques are crucial in filtering out noise, identifying trends, and making accurate predictions. For example, processing data from a network of air quality sensors can provide real-time alerts and long-term insights into pollution patterns, informing public health policies and community actions to mitigate environmental hazards.

Public safety is also significantly enhanced through innovations in signal processing. Surveillance systems employing video and audio signal processing are pivotal in crime prevention and incident response. For instance, advanced video analytics can automatically detect suspicious activities, recognize faces, and track movements in real-time, thereby augmenting law enforcement capabilities. Additionally, acoustic signal processing can pinpoint the location of gunshots or other distress sounds, enabling faster emergency response and potentially saving lives.

In the social domain, signal processing initiatives contribute to bridging the digital divide and fostering inclusive communication. Speech recognition and natural language processing algorithms enable the creation of accessible technologies for individuals with disabilities. For example, devices that convert spoken language into text help individuals with hearing impairments, while text-to-speech technologies provide reading assistance for those with visual impairments. These innovations not only enhance day-to-day living for many but also promote equal opportunities and social participation.

Moreover, signal processing is increasingly applied in developing sophisticated systems that facilitate education and social learning. For example, educational platforms use signal processing to analyze student interactions and behaviors, providing personalized learning experiences and identifying students who might need additional support. This approach helps in tailoring educational content and fostering a more inclusive and effective learning environment.

Furthermore, advances in signal processing support the development of early warning systems for natural disasters, providing communities with critical time to prepare and evacuate. By analyzing seismic data, weather patterns, and other environmental signals, these systems can predict events such as earthquakes, tsunamis, and hurricanes, significantly reducing the potential for loss of life and property.

In conclusion, the application of signal processing to social issues exemplifies the profound impact that technical innovation can have on society. By transforming data into actionable insights and developing robust, real-time solutions, signal processing initiatives address critical challenges in healthcare, environmental protection, public safety, social inclusion, and education. As technology continues to evolve, the potential for signal processing to contribute to a safer, healthier, and more equitable world will only expand, highlighting the importance of interdisciplinary collaboration and continued research in this dynamic field.

13.3 Measuring the Social Impact of Signal Processing Strategies

In today's interconnected world, signal processing plays an increasingly crucial role in various technological domains, from telecommunications to healthcare, entertainment, and environmental monitoring. Evaluating the social impact of these signal processing strategies requires a multidimensional approach that goes beyond mere technical performance. This involves considering the ethical, economic, and socio-cultural implications of deploying these technologies. In the following discussion, we will delve into how to measure this social impact, examining key factors and methodologies.

First and foremost, an essential component of measuring social impact lies in its ethical dimension. Signal processing technologies can have profound implications on privacy and surveillance. Consider, for instance, facial recognition systems integrated within public spaces. While they can enhance security by identifying potential threats, they also raise significant privacy concerns. To measure this impact, ethical audits can be conducted where various stakeholders—ranging from citizens to policy makers—are invited to discuss and outline the ethical boundaries and expectations from such technologies. Surveys and public consultations can elucidate public sentiment, thereby providing a more comprehensive understanding of ethical considerations.

Moreover, economic impact is another pivotal area. Signal processing strategies inevitably reshape industries by introducing efficiency and enabling new services. For example, advanced algorithms that compress data more efficiently can revolutionize data storage and transmission industries, reducing costs and enhancing service delivery. To quantify this impact, economic models can be constructed using key performance indicators (KPIs) such as cost savings, revenue growth, and job creation or loss attributable to the technology. Economic impact assessments can be carried out through these models, often involving rigorous data collection and analysis. Moreover, regulatory frameworks and economic policies may need to be considered, as they significantly influence the deployment and scalability of these strategies.

Community engagement is another crucial aspect that should not be overlooked. The success and social acceptance of signal processing strategies often hinge on how well they resonate with the community's needs and values. Participatory approaches, such as community-based participatory research (CBPR), can be effective in gathering insights and fostering a sense of ownership among community members. By involving local communities in the development, implementation, and evaluation phases, it is possible to ensure that the strategies are tailored to the community's specific socio-cultural context. Measures of social capital, such as trust, cooperation, and the strength of community networks, can serve as indicators of social impact. Surveys and interviews can provide qualitative and quantitative data that capture the nuances of community perspectives.

The educational implications of signal processing strategies are also important. Advanced signal processing can democratize access to information and education, breaking down geographical and socio-economic barriers. For instance, the use of adaptive streaming technologies in e-learning platforms enables users with limited bandwidth to access educational content without significant degradation in quality. Evaluating the social impact in this context can involve metrics such as student engagement levels, learning outcomes, and the accessibility of educational content across different demographics. Longitudinal studies can be particularly valuable, offering insights into how these technologies influence educational trajectories over time.

Health and well-being present another significant dimension. Signal processing technologies in healthcare, such as those used in medical imaging and telemedicine, can drastically improve patient outcomes and access to services. Evaluating the social impact here may involve metrics related to health outcomes—such as morbidity and mortality rates, patient satisfaction, and the accessibility of services. Additionally, considerations around equity and inclusivity are crucial. Are these technologies equally accessible to all socio-economic groups? Are there disparities in the benefits received? Quantitative data from healthcare providers, along with qualitative data from patient interviews, can provide a holistic view of the impact.

Lastly, environmental sustainability is an increasingly critical factor. Signal processing technologies can contribute to more efficient resource utilization and waste reduction, particularly in industries like agriculture and energy. The social impact in this domain can be measured through environmental KPIs such as carbon footprint reduction, energy efficiency gains, and waste minimization. Life Cycle Assessment (LCA) methodologies can also provide a comprehensive evaluation of the environmental impact over the entire lifespan of the technology.

In conclusion, measuring the social impact of signal processing strategies requires a multifaceted approach that addresses ethical, economic, socio-cultural, educational, health, and environmental dimensions. Employing a combination of qualitative and quantitative methodologies will yield the most comprehensive insights. This holistic evaluation can ensure that these technologies are developed and implemented in ways that are socially beneficial and aligned with broader societal goals.

13.4 Case Studies in Socially Responsible Signal Processing

In recent years, the field of signal processing has witnessed a significant transformation, driven by an increasing emphasis on socially responsible practices. This emerging trend underscores the necessity of integrating ethical considerations, societal impact, and environmental sustainability into the technical rigor of signal processing. By examining select case studies, one can appreciate how these principles are being interwoven into the fabric of signal processing in diverse applications.

One notable case is in the realm of healthcare, where socially responsible signal processing has made profound contributions. For instance, the development of wearable health monitoring devices has revolutionized patient care, particularly for individuals with chronic conditions such as diabetes and cardiovascular diseases. These devices employ advanced signal processing algorithms to continuously monitor vital signs like heart rate, blood glucose levels, and blood pressure. The data, processed in real-time, not only affords patients greater autonomy but also enables timely interventions by healthcare providers. This proactive approach not only improves patient outcomes but also reduces the burden on healthcare systems, thus exemplifying how signal processing can yield significant social benefits.

Beyond healthcare, environmental sustainability presents another compelling area where signal processing is making a meaningful impact. Consider the deployment of sensor networks for environmental monitoring. These networks, powered by sophisticated signal processing techniques, can detect and analyze a plethora of environmental parameters such as air and water quality, noise pollution, and soil moisture levels. For example, in the context of urban air quality monitoring, signal processing algorithms can discern patterns and identify sources of pollution with high precision. This data, when shared with policymakers and the public, can drive initiatives to curb pollution and promote cleaner urban environments, demonstrating the role of signal processing in fostering a healthier planet.

Education, too, provides fertile ground for the application of socially responsible signal processing. The ongoing evolution of education technology has brought about adaptive learning systems that cater to the diverse needs of students. These systems leverage signal processing to analyze data from various inputs—such as student interaction patterns, performance metrics, and even eye-tracking data—to create personalized learning experiences. By attending to individual learning styles and paces, these systems help bridge educational gaps, ensuring that all students, regardless of their background, have the opportunity to achieve their full potential. This approach underscores how signal processing can be harnessed to create more equitable educational outcomes.

Moreover, the applications of socially responsible signal processing extend to the realm of emergency response. The increased frequency of natural disasters calls for advanced systems capable of early detection and swift response. Signal processing plays a pivotal role in developing these systems; for instance, seismographs equipped with advanced algorithms can predict earthquakes by analyzing minute changes in ground vibrations. Similarly, weather radar systems use signal processing to provide accurate forecasts and warnings for extreme weather events like hurricanes and tornadoes. By enabling timely and coordinated response efforts, these technologies mitigate the impacts of disasters, save lives, and reduce economic losses.

The underlying principles of fairness, inclusivity, and sustainability also guide the design and implementation of signal processing systems. Researchers and practitioners are increasingly recognizing the importance of addressing biases in signal processing models, particularly those employed in critical domains like law enforcement and financial services. Developing algorithms that are transparent, interpretable, and devoid of bias is essential to ensuring that these systems serve all segments of society equitably.

In conclusion, the fusion of signal processing with socially responsible practices marks a pivotal evolution in the field. Through case studies across healthcare, environmental sustainability, education, and emergency response, it is evident how signal processing not only advances technological frontiers but also addresses pressing societal challenges. As this interdisciplinary approach continues to mature, it holds the promise of fostering more ethical, inclusive, and resilient communities.

14. The Future of Signal Processing

As we stand at the precipice of an era characterized by unprecedented technological advancement, the field of signal processing is poised to undergo transformative changes. Signal processing, the manipulation and analysis of signals to modify or improve their information content, has already revolutionized industries ranging from telecommunications to healthcare. Yet, the horizon promises even greater evolution driven by rapid developments in artificial intelligence, machine learning, and quantum computing.

In this chapter, we will explore the emerging trends and breakthroughs that are set to redefine the landscape of signal processing. From the integration of neural networks for more sophisticated pattern recognition, to advancements in real-time data processing that can handle the ever-growing influx of Big Data, the future is brimming with possibilities. We will examine not only the technological innovations but also the potential societal impacts, ethical considerations, and the new challenges that will arise in this dynamic field.

Join us as we delve into what lies ahead, envisioning a world where signal processing not only enhances human capabilities but also shapes the very fabric of our everyday lives.

14.1 Emerging Trends in Signal Processing Practices

The field of signal processing continues to evolve rapidly, driven by advancements in technology and the increasing complexity of data-driven applications. Emerging trends in signal processing practices reflect the shift towards more sophisticated, dynamic, and intelligent systems designed to tackle real-world challenges. Several key trends have gained prominence, influencing research directions and practical applications alike.

One significant trend is the integration of machine learning (ML) and artificial intelligence (AI) into signal processing frameworks. Traditional signal processing relied heavily on deterministic models and structured approaches to interpret and manipulate signals. However, the capability of ML and AI algorithms to learn from data and adaptively improve their performance has revolutionized the field. Techniques such as deep learning have shown tremendous promise in areas like image and speech recognition, where vast amounts of data and complex patterns are involved. Convolutional neural networks (CNNs) and recurrent neural networks (RNNs) are especially noteworthy, being employed extensively in processing visual and temporal signals, respectively. This integration has led to more robust, accurate, and efficient signal analysis and synthesis processes.

Another notable trend is the application of big data analytics to signal processing. The ubiquity of sensors and the proliferation of Internet of Things (IoT) devices have resulted in an explosion of data, necessitating scalable methods to handle, process, and extract meaningful information. Signal processing now needs to manage high-dimensional, high-velocity, and high-variety datasets. Big data techniques, such as distributed computing frameworks (e.g., Hadoop, Spark) and advanced data storage solutions (e.g., NoSQL databases), are being adapted to process signal data efficiently at scale.

The trend towards real-time and edge processing is also gaining traction. With the demand for instantaneous decision-making in applications like autonomous driving, industrial automation, and smart healthcare, there is a pressing need for signal processing systems that can operate with minimal latency. Edge computing addresses this requirement by processing data closer to the source, reducing the dependency on central cloud infrastructures and improving response times. Real-time processing algorithms and hardware accelerators like field-programmable gate arrays (FPGAs) and graphics processing units (GPUs) are critical enablers of this trend, ensuring that signal processing tasks are performed swiftly and efficiently.

In the realm of communications, the advent of 5G technology has ushered in a new era of signal processing challenges and opportunities. With its promises of high data rates, low latency, and massive connectivity, 5G necessitates advanced signal processing techniques to manage the increased complexity of wireless communication systems. Techniques such as massive multiple-input multiple-output (MIMO) and millimeter-wave communications require sophisticated signal processing algorithms for efficient beamforming, interference management, and channel estimation. These advancements are set to revolutionize not just telecommunications but also various industries relying on high-speed, reliable connectivity.

Quantum signal processing is another cutting-edge area attracting considerable interest. Quantum computing's potential to solve certain problems more efficiently than classical computing has significant implications for signal processing. Quantum algorithms can, in theory, provide exponential speedups for tasks such as Fourier transforms, which are fundamental to signal processing. While still in its nascent stages, research in quantum signal processing looks promising, with potential applications ranging from cryptography to complex system simulations.

Adaptive and personalized signal processing is also a growing trend. This approach tailors signal processing strategies to individual user preferences or specific environmental conditions. For instance, personalized hearing aids that adapt to differing sound environments in real-time or adaptive learning systems that modify content delivery based on a student's progress exemplify this trend. The ability to dynamically adjust processing in response to contextual changes significantly enhances system effectiveness and user satisfaction.

Finally, interdisciplinary collaborations are fostering new innovations in signal processing. By integrating insights from fields such as biology, neuroscience, and materials science, researchers are developing novel signal processing methods inspired by natural systems. Bio-inspired algorithms, for instance, mimic neural networks found in the human brain, offering new ways to process and interpret signals.

In summary, emerging trends in signal processing practices are being shaped by technological advancements and the growing demands of modern applications. The incorporation of machine learning, big data techniques, real-time processing, 5G communications, quantum computing, adaptive systems, and interdisciplinary approaches are all propelling the field forward. These developments hold the promise of more intelligent, efficient, and versatile signal processing solutions, capable of addressing the complexities of today's data-driven world.

14.2 Globalization and Its Impact on Signal Processing

Globalization has markedly influenced numerous scientific and engineering fields, and signal processing is no exception. The intersection of these two domains has led to transformative changes in the way information is acquired, processed, transmitted, and utilized worldwide. Signal processing, which involves the analysis, modification, and synthesis of signals such as sound, images, and scientific measurements, has benefited immensely from the global interconnectedness brought about by globalization.

Firstly, the ubiquitous nature of the internet, a quintessential product of globalization, has revolutionized signal processing. It has facilitated unprecedented access to data and computational resources. Researchers and engineers around the globe can now share algorithms, data sets, and findings instantaneously. This democratization of information enables collaborative efforts that were once constrained by geographic boundaries. Consequently, advancements in signal processing techniques can be shared rapidly, accelerating innovation and improving solutions to complex problems in real-time.

The extensive proliferation of digital devices globally adds another layer to the impact of globalization on signal processing. Mobile phones, tablets, and other digital gadgets generate vast amounts of data every second. This colossal influx of data necessitates sophisticated signal processing techniques to filter, compress, and interpret the signals accurately. Here, globalization plays a pivotal role by ensuring a global standardization of technologies. Standards such as 4G, 5G, Wi-Fi, and others ensure that signal processing algorithms are universally applicable, thus enhancing interoperability and easing communication across borders.

In addition, the globalization of markets has led to a heightened demand for products and services that are both efficient and cost-effective. This demand has spurred innovation in hardware and software solutions associated with signal processing. For instance, the development of powerful, yet affordable processing units like GPUs (Graphics Processing Units) enables complex signal processing tasks to be executed faster and more efficiently. Similarly, globalization has fostered competitive markets which drive down the costs of such technologies, making advanced signal processing capabilities accessible to a broader audience.

Cultural exchange, facilitated by globalization, has also impacted signal processing. Different regions often have unique requirements and constraints based on their specific socio-economic and cultural contexts. For example, certain signal processing applications in telecommunication or multimedia may need to accommodate distinct languages, dialects, and cultural idiosyncrasies. The global exchange of ideas and expertise enables the development of versatile and adaptive algorithms that cater to diverse needs, thus ensuring that signal processing technologies are inclusive and globally relevant.

Moreover, the educational landscape has been transformed by globalization, significantly benefiting the field of signal processing. Access to global educational resources, online courses, and international conferences allows budding engineers and seasoned professionals alike to keep abreast of the latest developments. The cross-pollination of ideas fostered by global academic and professional networks ensures continuous evolution and refinement of signal processing methodologies.

Another critical impact of globalization on signal processing is the accelerated pace of technological convergence. Technologies such as artificial intelligence (AI), machine learning (ML), and big data analytics increasingly intertwine with signal processing. Global collaborations and research initiatives have catalyzed this convergence, leading to more sophisticated and intelligent signal processing techniques. AI-driven signal processing, for instance, optimizes data interpretation and decision-making processes across various applications, from medical diagnostics to autonomous systems.

Despite these positive impacts, globalization also presents certain challenges to the field of signal processing. The increased interconnectivity and data sharing heighten the risk of cybersecurity threats. Signal processing techniques must therefore continuously evolve to safeguard data integrity and privacy. Furthermore, global standardization processes can sometimes stymie innovation by imposing rigid frameworks that may not account for regional specificities or emerging technologies.

In conclusion, globalization has undeniably reshaped the landscape of signal processing, bringing about significant advancements and facilitating global collaboration. Enhanced access to data and resources, the ubiquity of digital devices, competitive markets, cultural exchange, and educational enrichment collectively propel the field forward. However, these advancements must be balanced with robust cybersecurity measures and flexible standardization to fully harness the potential of globalization in signal processing.

14.3 Navigating Technological Uncertainty with Signal Processing Insights

In an era characterized by rapid technological advancements and constant innovation, organizations and individuals alike are often confronted with considerable uncertainty about the future. This uncertainty stems from various sources, including shifts in market demands, disruptions precipitated by emerging technologies, and the evolving landscape of regulatory requirements. One effective way to mitigate and navigate this technological uncertainty is through the application of signal processing insights.

Signal processing, fundamentally, involves the analysis, interpretation, and manipulation of signals to extract meaningful information. Though traditionally applied in the realms of electrical engineering and communications, the principles of signal processing can be extended to a broader technological context. Its core techniques, such as filtering, transformation, and statistical analysis, prove invaluable in distilling clarity from chaos.

A primary aspect of signal processing is noise reduction. In the technological domain, noise can be equated to irrelevant data or market 'buzz' that obscures critical information. Advanced filtering techniques, such as Kalman filters or wavelet transforms, can be employed to isolate significant trends from background noise. For instance, businesses can apply these techniques to social media data, discerning authentic consumer sentiments from the cacophony of online chatter. By doing so, decision-makers can make more informed judgments regarding product development, marketing strategies, and customer engagement efforts.

Another pivotal contribution of signal processing is in the realm of predictive analytics. Techniques such as Fourier transforms and autoregressive models allow for the decomposition of historical data to identify underlying patterns and periodicities. This can be especially useful for forecasting future technological trajectories or market behaviors. For example, in the semiconductor industry, where rapid obsolescence is a constant concern, understanding cyclical trends in innovation and production can help companies better predict future product lifecycles and invest accordingly.

Moreover, advanced machine learning algorithms, which are intrinsically connected to the principles of signal processing, are becoming essential tools in the arsenal against technological uncertainty. Machine learning models, like neural networks and support vector machines, are adept at recognizing patterns within vast and complex datasets. These models can predict market disruptions by identifying subtle shifts in data that may precede major technological breakthroughs or failures. Google's AlphaGo, which integrated deep learning with signal processing techniques, exemplifies the power of these methods. It demonstrated an ability to predict and adapt to complex scenarios, offering a blueprint for navigating technological complexity.

Real-time signal processing is another aspect that offers significant benefits. In industries such as autonomous vehicles or cybersecurity, the ability to process and react to signals instantaneously can be crucial. For autonomous vehicles, real-time signal processing can interpret sensory data from LIDAR and cameras to make split-second decisions, enhancing safety and performance. In cybersecurity, advanced signal processing algorithms can detect anomalies in network traffic, preempting potential security breaches.

Furthermore, the democratization of signal processing tools and techniques is making them accessible to a broader audience beyond traditional engineering disciplines. Open-source platforms and software libraries, such as MATLAB, NumPy, and TensorFlow, provide robust environments for implementing and experimenting with signal processing algorithms. This availability allows businesses in diverse fields - from finance to healthcare - to leverage signal processing insights without necessitating specialized expertise.

The interdisciplinary nature of signal processing also facilitates collaborative innovation. By integrating insights from fields like artificial intelligence, big data analytics, and operational research, organizations can develop comprehensive strategies to navigate technological uncertainty. This holistic approach ensures that decisions are not made in isolation but are informed by a synthesis of diverse expertise and perspectives.

In conclusion, the application of signal processing insights offers a pragmatic and powerful approach to navigating the uncertainties inherent in modern technological landscapes. By effectively filtering out noise, predicting trends, enabling real-time decision-making, and fostering interdisciplinary collaboration, signal processing empowers organizations to anticipate and adapt to change with greater confidence and agility. As technological ecosystems continue to evolve, the strategic implementation of these insights will be indispensable in shaping resilient and forward-thinking enterprises.

14.4 Preparing for the Future of Signal Processing

Embracing the future of signal processing necessitates a keen understanding of both the current trends and the evolving landscape of technological advancements. As our world becomes ever more connected and complex, signal processing must adapt to meet new challenges and harness emerging opportunities. This preparation hinges on several key areas: education and training, research and development, integration of artificial intelligence, advancements in hardware, and the ethical considerations of these advancements.

Education and Training: To stay at the forefront of signal processing, it's imperative to invest in comprehensive education and continuous professional development. Curricula in universities need to incorporate the latest advancements in the field, such as machine learning, deep learning, and quantum computing. By doing so, they can ensure that graduates are well-equipped with the knowledge and skills required to innovate and excel. Furthermore, professionals in the industry must engage in continuous learning opportunities through workshops, conferences, and online courses to stay updated with the latest tools and techniques.

Research and Development: Research is the cornerstone of progress in signal processing. Future advancements will likely stem from interdisciplinary collaborations, where experts from fields such as data science, neuroscience, and bioengineering contribute to groundbreaking developments. Funding and support for both basic and applied research are essential. Governments, academic institutions, and private sectors need to foster environments where innovative ideas can germinate and mature into practical applications.

Integration of Artificial Intelligence: Artificial intelligence (AI) is poised to revolutionize signal processing. AI algorithms can enhance signal processing tasks such as filtering, noise reduction, and pattern recognition. Machine learning models, particularly deep learning networks, exhibit remarkable capabilities in interpreting complex data. For instance, convolutional neural networks (CNNs) have shown exceptional performance in image and speech recognition tasks. The integration of AI can lead to more robust and adaptive signal processing systems that learn and improve over time.

Advancements in Hardware: The evolution of hardware technology is a critical factor in the future of signal processing. The development of more powerful and energy-efficient processors, such as graphics processing units (GPUs) and specialized neural network accelerators, enables the handling of large-scale data necessary for real-time processing. Moreover, the advent of quantum computing promises to solve problems currently deemed intractable by classical computers. Researchers are exploring how quantum algorithms can be applied to signal processing to achieve unprecedented computational speeds and accuracy.

Ethical Considerations: As signal processing technology advances, it engenders new ethical dilemmas. Ensuring data privacy and security is paramount, especially as signal processing applications often deal with personal and sensitive information. Ethical frameworks must be established to guide the development and deployment of these technologies. Transparency in AI-driven signal processing systems is crucial to maintain public trust. Additionally, there is a need to consider the societal impacts of automation driven by advanced signal processing techniques, including potential job displacement and the digital divide.

In preparing for the future, stakeholders in signal processing must also consider the environmental footprint of burgeoning technologies. Sustainable practices should be incorporated into the design and manufacturing of signal processing hardware. This includes optimizing energy efficiency and exploring alternative materials that reduce environmental impact.

Collaboration across disciplines and sectors will be instrumental in navigating the future of signal processing. Academia, industry, and policy-makers need to work in concert to address the challenges and exploit the opportunities on the horizon. Such cooperation can accelerate innovation, drive economic growth, and ultimately contribute to the betterment of society.

In conclusion, the future of signal processing is a dynamic and multifaceted domain poised for significant evolution. Preparing for this future requires a concerted effort in education, research, and ethical governance. By fostering innovation and collaboration, we can pave the way for advancements that not only enhance technological capabilities but also uphold societal values and sustainability.

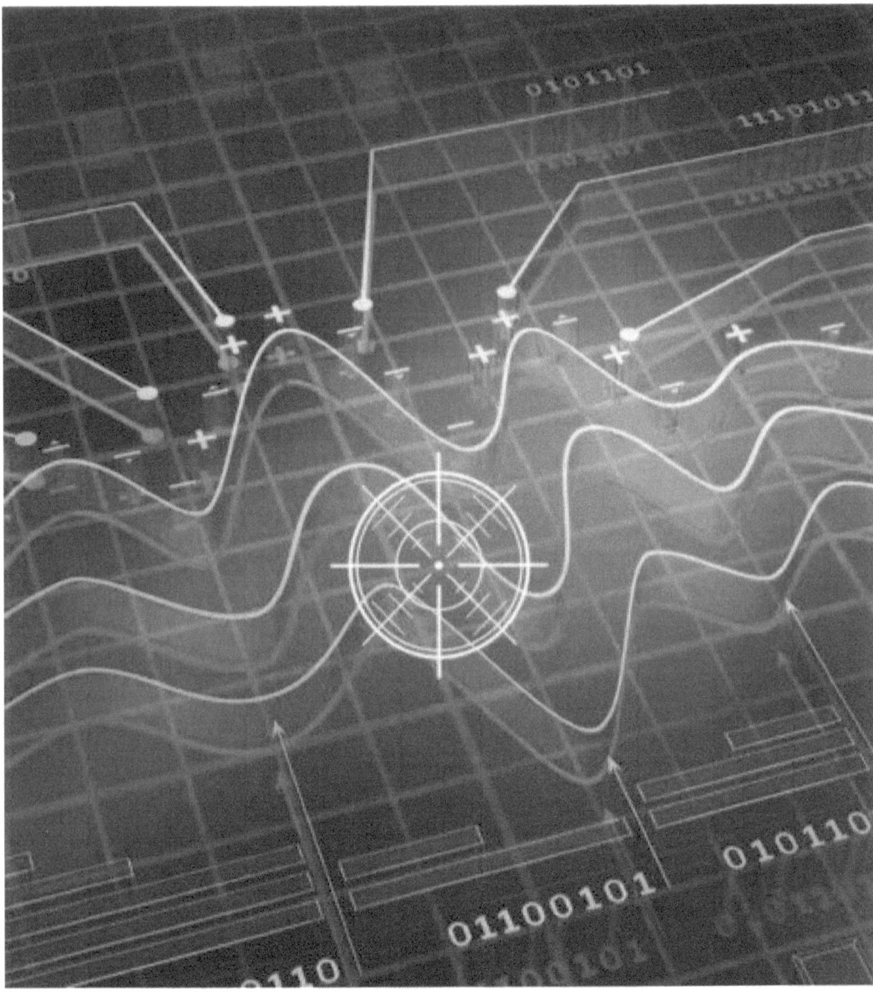
01100101

15. Scaling Signal Processing Practices for Success

Engineering successful signal processing systems requires more than just developing sophisticated algorithms and deploying them on efficient hardware. As our systems grow in complexity and our datasets increase in size, it becomes crucial to scale our practices effectively. Scaling involves refining methodologies, optimizing computational resources, and ensuring the robustness of algorithms across diverse scenarios and volumes of data. This chapter delves into the strategic principles and practical techniques that can empower your signal processing systems to adapt and thrive under the pressures of growth and expanded demand. From leveraging parallelism to enhance computational efficiency, to employing advanced machine learning models for adaptive signal processing, we will explore an array of strategies designed to elevate your operations. By building a strong foundation for scalability, not only can we handle today's challenges, but we also set the stage for future innovation and success.

15.1 Strategies for Scaling Signal Processing Across Systems

In modern engineering, signal processing plays a pivotal role in various applications, from telecommunications to medical diagnostics and beyond. With advancements in technology and the growing complexity of applications, scaling signal processing across systems has become a critical challenge. Effective scaling ensures that systems can handle increasing amounts of data while maintaining performance and reliability. This necessitates the implementation of robust strategies tailored to manage computational demands and resource constraints.

The first strategy to consider is parallel processing. By dividing the signal processing tasks among multiple processors, a system can significantly increase its throughput. This approach is particularly useful in applications where real-time processing is essential. For example, in a radar system, parallel processing can be employed to manage multiple target tracking, where each processor handles a subset of the incoming signals. Such distribution not only accelerates processing but also enhances fault tolerance, as the failure of one processor does not cripple the entire system.

Another valuable strategy is leveraging specialized hardware, such as Digital Signal Processors (DSPs) and Field-Programmable Gate Arrays (FPGAs). DSPs are designed specifically for high-speed numeric operations, making them ideal for tasks like filtering and Fourier transforms. FPGAs offer flexibility, allowing developers to create custom hardware accelerators tailored to specific signal processing algorithms. Both DSPs and FPGAs can offload intensive computations from general-purpose processors, thus freeing them to handle other tasks and improving overall system efficiency.

The third strategy revolves around optimized algorithm design. Algorithms that are computationally efficient can dramatically reduce the load on processing units. For instance, implementing algorithms with lower complexity, such as using Fast Fourier Transform (FFT) instead of a naive Discrete Fourier Transform (DFT), can cut down processing time and resource utilization. Furthermore, algorithms should be designed with concurrency in mind, ensuring that they can be efficiently decomposed into parallelizable tasks.

Distributed computing is another powerful strategy, wherein signal processing tasks are spread across multiple systems in a networked environment. This approach leverages the collective power of multiple machines, often leading to near-linear scalability. Cloud computing platforms provide an excellent infrastructure for distributed computing, offering virtually unlimited resources on-demand. A cloud-based approach also benefits from elasticity, allowing systems to scale up or down dynamically based on current workload.

Software frameworks and libraries optimized for signal processing can also play a significant role. For instance, frameworks like CUDA and OpenCL enable signal processing applications to harness the power of GPU acceleration. These libraries provide developers with tools to write highly parallelized code that can run on multiple hardware architectures. Utilizing such frameworks not only simplifies development but also maximizes hardware utilization, leading to more scalable systems.

Resource management is another crucial aspect. Effective scaling necessitates careful planning and allocation of computational resources. Techniques such as load balancing and dynamic resource allocation ensure that no single processor is overwhelmed, and resources are used optimally. In multicore systems, affinity scheduling can bind specific tasks to designated cores, improving cache utilization and minimizing inter-core communication overhead.

Moreover, adopting hierarchical processing models can enhance scalability. In such models, signal processing tasks are structured in a hierarchical manner, where high-level tasks supervise and distribute work to lower-level tasks. This hierarchical approach mirrors biological neural networks, leading to highly scalable and efficient processing.

Lastly, scalability metrics and monitoring tools are indispensable. Continuous monitoring of system performance helps identify bottlenecks and optimize resource allocation. Tools like performance profilers and log analyzers provide insights into system behavior under different loads, guiding developers to fine-tune their systems.

In summary, scaling signal processing across systems requires a combination of parallel processing, specialized hardware, optimized algorithms, distributed computing, software frameworks, effective resource management, hierarchical processing models, and robust monitoring tools. By combining these strategies, engineers can design systems that are not only capable of handling increased data volumes but also adaptable to future advancements and demands.

15.2 Managing Signal Processing in High-Demand Environments

In today's rapidly evolving technological landscape, managing signal processing in high-demand environments is a critical challenge that requires a multifaceted approach. Signal processing, the manipulation and analysis of signals to optimize their transmission, storage, and quality, becomes increasingly complex under high-demand conditions. These environments, which can range from telecommunications networks to radar systems, demand robust strategies to ensure efficiency, reliability, and real-time performance.

One of the foremost considerations in these settings is the optimization of computational resources. Signal processing algorithms, such as Fast Fourier Transforms (FFT) or digital filtering, are computationally intensive and can strain the processing capabilities of the hardware. Therefore, the selection of efficient algorithms is paramount. This involves choosing algorithms that not only perform the required tasks but do so with minimized computational load and latency. Implementing multithreaded or parallel processing techniques can significantly enhance performance, leveraging modern multi-core processors to handle multiple signal streams simultaneously.

In high-demand environments, data throughput and latency are crucial metrics. The bandwidth of the transmission medium and the speed of data processing directly influence the overall system performance. Techniques like data compression can be beneficial, reducing the amount of data that needs to be processed and transmitted without compromising the integrity of the signal. Additionally, error detection and correction methods must be deployed to ensure data integrity. Algorithms such as Reed-Solomon or convolutional codes can detect and correct errors in real-time, which is essential for maintaining the fidelity of signals in noisy or unpredictable environments.

Another significant aspect is the adaptive management of processing loads. High-demand environments often exhibit fluctuating demands which can overwhelm the system if not managed adeptly. Dynamic load balancing techniques can be employed to distribute the processing load evenly across the available resources. This involves real-time monitoring and analysis of resource utilization, followed by adjustments to the allocation of tasks. Techniques such as adaptive filtering can also be used, where the parameters of the filter are adjusted in real-time based on the changing characteristics of the input signal.

Network topology and architecture play a pivotal role in signal processing efficiency. Designing a scalable and resilient network architecture ensures that as demand increases, the system can expand without significant degradation in performance. Incorporating redundancy and failover mechanisms enhances reliability, ensuring continuous operation even in the event of hardware failures or unexpected surges in demand. Further, employing distributed processing models, where processing tasks are spread across multiple nodes, can alleviate the pressure on any single component and enhance the overall resilience of the system.

Signal processing in high-demand environments also necessitates robust security measures. The transmission and processing of signals often involve sensitive information, making security a top priority. Encryption algorithms, secure communication protocols, and rigorous access controls are essential to protect data integrity and confidentiality. Additionally, anomaly detection systems can identify potential security breaches or abnormal behaviors in real-time, enabling prompt response to mitigate risks.

Real-time monitoring and analytics are indispensable tools in managing high-demand environments. By continuously collecting and analyzing performance metrics, operators can gain insights into system behavior and identify potential bottlenecks. Advanced machine learning algorithms can predict future demand patterns and optimize resource allocation proactively. These analytics can also facilitate preventive maintenance, identifying signs of potential hardware or software failures before they occur.

In conclusion, the management of signal processing in high-demand environments requires a holistic approach that integrates efficient algorithms, adaptive load management, robust network architecture, stringent security measures, and real-time analytics. By addressing these areas comprehensively, it is possible to maintain high levels of performance, reliability, and security, ensuring that the system can meet the demands of modern technological applications. As the complexity and scale of these environments continue to grow, ongoing innovation and refinement in signal processing techniques will remain essential.

15.3 Expanding Signal Processing Capabilities Globally

Signal processing has evolved significantly over the past few decades, transforming numerous industries by enhancing data interpretation, communication, and multimedia experiences. As technology continues to advance, the capability to expand signal processing on a global scale is becoming increasingly vital.

One of the primary drivers for the global expansion of signal processing capabilities is the ever-growing volume of data generated across various sectors. From telecommunications and healthcare to finance and entertainment, the need to efficiently process and analyze large datasets cannot be overstated. Signal processing techniques enable the extraction of meaningful information from raw data, aiding in decision-making processes, enhancing system performance, and improving user experiences.

Technological advancements in hardware and software play a crucial role in this expansion. The rise of powerful processors, coupled with sophisticated algorithms, has paved the way for real-time signal processing applications. In the past, the computational demands of such tasks were often a limiting factor. However, with the advent of multi-core processors, graphics processing units (GPUs), and specialized chips designed for signal processing, these limitations have been significantly reduced.

Software advancements, particularly in the field of artificial intelligence (AI) and machine learning (ML), have further augmented signal processing capabilities. AI and ML algorithms can be trained to recognize patterns and make predictions based on signal data, which is particularly useful in applications such as image and speech recognition, medical diagnostics, and predictive maintenance in industrial settings. By leveraging AI and ML, signal processing systems can become more adaptive and intelligent, providing superior accuracy and efficiency.

Moreover, the integration of wireless communication technologies plays an essential role in broadening the reach of signal processing solutions. The proliferation of 4G and 5G networks has not only improved data transmission speeds but also enhanced connectivity in remote and underserved areas. This connectivity is instrumental in implementing distributed signal processing systems, where data is collected, processed, and transmitted across various nodes in a network. Applications such as the Internet of Things (IoT) heavily rely on such capabilities, enabling smart homes, cities, and industries.

Collaborative efforts on an international scale are also critical in this expansion. Research institutions, corporations, and governments around the world are increasingly recognizing the importance of signal processing. Initiatives aimed at fostering innovation and knowledge sharing are essential in addressing global challenges. Conferences, workshops, and symposia dedicated to signal processing provide platforms for researchers and practitioners to exchange ideas, present findings, and collaborate on groundbreaking projects.

Standardization plays an essential role in facilitating global expansion. By establishing common protocols and standards, interoperability between different systems is ensured, promoting the seamless integration of signal processing technologies across borders. Organizations such as the Institute of Electrical and Electronics Engineers (IEEE) and the International Telecommunication Union (ITU) are actively involved in developing standards that guide the implementation and deployment of signal processing solutions worldwide.

The educational aspect is also worth noting. As the demand for skilled professionals in signal processing continues to grow, educational institutions around the globe are expanding their curricula to include advanced courses on this subject. Online learning platforms are making these courses accessible to a wider audience, thereby democratizing knowledge and fostering a new generation of experts capable of driving the field forward.

Lastly, addressing ethical considerations and ensuring data privacy and security are paramount when expanding signal processing capabilities globally. As data becomes more intertwined with personal and sensitive information, maintaining the integrity and confidentiality of this data is crucial. Strategies such as encryption, anonymization, and robust cybersecurity measures must be implemented to protect against data breaches and misuse.

In summary, the global expansion of signal processing capabilities is a multifaceted endeavor that hinges on technological advancements, international collaboration, standardization, education, and ethical practices. As we progress further into the digital age, the ability to efficiently process and analyze signals will be indispensable in solving complex problems and driving innovation across various domains.

15.4 Innovative Approaches to Scaling Signal Processing

Scaling signal processing presents complex challenges, necessitating innovative approaches to address evolving data demands. The emergence of big data, increasing sensor networks, and more sophisticated communication systems have highlighted the need for more advanced methodologies in this domain. This discourse will delve into three promising innovative approaches: adaptive algorithms, machine learning integration, and quantum computing.

First, adaptive algorithms have revolutionized the scaling of signal processing. These algorithms adjust their parameters in real-time based on input data characteristics and system performance, thereby optimizing processing efficiency. One key aspect lies in their ability to balance computational loads dynamically. Traditional static algorithms do not efficiently handle variable workloads or changing data statistics. In contrast, adaptive algorithms can scale resources as needed, maximizing efficiency. For instance, stochastic gradient descent (SGD) and its variants have been widely adopted for adaptive filtering and optimization in signal processing tasks. This approach minimizes computational overhead by iteratively refining the model parameters based on incoming data, which is especially effective in environments where data patterns are non-stationary or evolving.

Secondly, the integration of machine learning (ML) techniques into signal processing workflows stands out as a transformative approach. ML algorithms, particularly deep learning models, have consistently demonstrated superior performance in pattern recognition, feature extraction, and classification tasks. By leveraging neural networks, signal processing applications can benefit from automatic feature learning, reducing the need for manual feature engineering. One significant advantage of integrating ML with signal processing is its ability to handle high-dimensional data efficiently. For example, convolutional neural networks (CNNs) have been successfully employed in image and speech recognition tasks, offering scalability and accuracy unmatched by traditional signal analysis techniques. Moreover, ML-driven signal processing systems can be trained to adapt and improve over time as more data becomes available, ensuring that the system scales effectively with growing datasets.

Furthermore, transfer learning techniques within the ML domain enable scalability by reusing pre-trained models on new but related tasks. This reduces the computational burden associated with training new models from scratch, thereby accelerating the deployment of scalable signal processing solutions. In practice, pre-trained models on vast datasets such as ImageNet have been repurposed for signal processing applications, including medical diagnostics, remote sensing, and communication systems, demonstrating significant improvements in efficiency and performance.

Lastly, quantum computing offers a forward-looking approach to scaling signal processing. By exploiting quantum mechanical properties such as superposition and entanglement, quantum computing promises to handle computations that are infeasible for classical computers. Quantum algorithms, like quantum Fourier transform (QFT) and quantum phase estimation, hold potential for accelerating signal processing tasks such as spectral analysis, data compression, and noise reduction. These quantum-based methods can, theoretically, process multiple states simultaneously, drastically reducing the time complexity of certain signal processing operations. For instance, QFT can efficiently compute the discrete Fourier transform (DFT), a fundamental operation in signal processing, which classical algorithms perform less efficiently as data size scales.

Additionally, hybrid approaches that combine quantum and classical computing could address current technological limitations and serve as a bridge towards fully quantum-based signal processing systems. Researchers are actively exploring these hybrid models to solve practical signal processing problems more efficiently than classical methods alone.

In conclusion, adaptive algorithms, machine learning integration, and quantum computing represent three innovative approaches to scaling signal processing. Each method addresses specific challenges associated with data size, complexity, and computational demands. Together, these approaches hold the promise of creating more efficient, scalable, and robust signal processing systems capable of meeting the demands of contemporary and future applications. As technology advances, leveraging these innovations will be crucial in maintaining the pace of progress in ever-expanding data environments.

16. Corporate Strategies in Signal Processing

In the modern landscape of technology and business, corporate strategies in the realm of signal processing have emerged as a pivotal factor for success and innovation. Signal processing, once the exclusive domain of researchers and specialized engineers, has transcended its traditional boundaries to become integral to various industries, ranging from telecommunications to healthcare, automotive to consumer electronics.

The significance of this field lies in its ability to transform raw data into meaningful information, enabling the development of sophisticated systems and solutions. With the exponential growth of data and the advent of advanced computational capabilities, corporations are increasingly recognizing the strategic importance of investing in signal processing technologies and expertise. This chapter delves into the diverse strategies that corporations adopt to harness the power of signal processing, examining how these strategies contribute to competitive advantage, operational efficiency, and market leadership.

We will explore the multifaceted approaches that companies take, including mergers and acquisitions, strategic partnerships, in-house research and development, and leveraging open innovation ecosystems. Through detailed case studies and analyses, this chapter aims to provide a comprehensive understanding of how corporate strategies in signal processing are formulated, implemented, and evolved in response to emerging challenges and opportunities. Whether you are a business leader, a technology enthusiast, or an academic researcher, this chapter offers valuable insights into the strategic imperatives driving this dynamic and critically important field.

16.1 Strategic Partnerships and Their Impact on Signal Processing

In today's rapidly evolving technological landscape, strategic partnerships play a vital role in advancing the field of signal processing. These collaborations, which can take various forms such as joint ventures, research alliances, and strategic agreements, foster innovation and drive technological advancements by pooling resources, expertise, and intellectual property. Understanding the impact of these partnerships on signal processing requires a detailed analysis of their contributions to research, development, and the overall growth of the field.

Firstly, strategic partnerships enhance research and development efforts in signal processing. By collaborating, organizations can combine their individual strengths and leverage shared expertise for more robust and comprehensive research initiatives. For example, university research labs often partner with technology companies to bridge the gap between theoretical research and practical application. These partnerships enable academic researchers to gain access to real-world data and industry-proven techniques, while companies benefit from cutting-edge research and innovative methodologies. As a result, both parties can accelerate the development of new signal processing algorithms and technologies, leading to more advanced and efficient solutions.

In addition to research, strategic partnerships also facilitate the sharing of resources, which is crucial for the advancement of signal processing. The development of new signal processing technologies often requires significant investments in terms of time, funding, and specialized equipment. By forming strategic alliances, organizations can share these costs and mitigate financial risks. For instance, two companies might jointly invest in the development of a new signal processing chip, thereby reducing the individual financial burden and allowing both to benefit from the eventual success of the technology. This collaborative approach enables organizations to undertake more ambitious projects than they could on their own, leading to faster and more impactful advancements in the field.

Furthermore, strategic partnerships can lead to the standardization of signal processing technologies. In order to ensure compatibility and interoperability between different systems and devices, it is essential to establish common standards and protocols. Collaboration between industry leaders, academic institutions, and standardization bodies can facilitate the development of these standards. For example, the development of the JPEG image compression standard involved collaboration between multiple stakeholders, including researchers, engineers, and industry representatives. By establishing a widely accepted standard, these partnerships help ensure that new signal processing technologies can be seamlessly integrated into existing systems, promoting widespread adoption and fostering further innovation.

Another significant impact of strategic partnerships on signal processing is the acceleration of technology transfer and commercialization. Bringing new technologies from the research stage to the market can be a complex and time-consuming process. Strategic partnerships between academic institutions, research organizations, and industry can streamline this process by providing the necessary infrastructure and support for commercialization. For instance, a university research lab might develop a novel signal processing algorithm, but lack the resources or expertise to bring it to market. By partnering with a technology company, the lab can leverage the company's manufacturing capabilities, marketing channels, and industry connections to commercialize the technology more effectively. This collaboration not only benefits the partners involved but also ensures that new signal processing advancements reach end users more quickly, enhancing the overall impact of the research.

Finally, strategic partnerships foster a culture of continuous learning and innovation within the field of signal processing. By working closely with partners from different sectors and disciplines, organizations can gain new insights and perspectives that drive further research and development. These collaborations often lead to the cross-pollination of ideas and the emergence of innovative solutions that might not have been possible in isolation. For example, a partnership between a healthcare provider and a technology company could result in the development of advanced signal processing techniques for medical imaging, combining the healthcare provider's domain expertise with the technology company's technical capabilities. This interdisciplinary approach not only enriches the field of signal processing but also ensures that it continues to evolve and address emerging challenges and opportunities.

In conclusion, strategic partnerships play a crucial role in advancing the field of signal processing by enhancing research and development efforts, sharing resources, standardizing technologies, accelerating technology transfer, and fostering a culture of innovation. These collaborations are essential for driving technological advancements and ensuring that new signal processing solutions can be developed and deployed effectively. As the field continues to evolve, the importance of strategic partnerships will only grow, making them a key factor in shaping the future of signal processing.

16.2 Mergers, Acquisitions, and Their Impact on Signal Processing

In recent years, the landscape of the signal processing industry has been significantly influenced by mergers and acquisitions (M&A). Mergers and acquisitions are potent corporate strategies wherein companies amalgamate through mutual agreement or one firm acquires another to consolidate resources, enhance capabilities, and achieve synergies. This phenomenon, although widespread in various industries, holds unique implications for the field of signal processing.

To understand the impact of M&A on signal processing, it is essential first to elucidate what signal processing entails. Signal processing involves the analysis, interpretation, and manipulation of signals. Signals can be audio, visual, or data streams, among others, which are processed to filter noise, enhance features, and extract useful information. Applications span across multiple fields, including telecommunications, medical imaging, radar systems, audio engineering, and more.

Mergers and acquisitions within this sector promote innovation and strategic growth due to the convergence of diverse technological assets and human expertise. For instance, when a firm specializing in cutting-edge digital signal processing algorithms acquires another entity proficient in hardware implementation, the resultant entity can leverage the combined proficiencies to develop integrated solutions that are both powerful and efficient. These consolidated resources can yield more comprehensive and innovative products, such as advanced signal processors and refined algorithms that enhance performance capabilities in end devices.

Moreover, the financial robustness attained post-M&A allows firms to invest more substantially in research and development (R&D). Increased R&D funding propels advancements in signal processing technologies. Enhanced R&D aligns with long-term advancements, enabling strides in artificial intelligence (AI) and machine learning (ML) applications within signal processing frameworks. These capabilities are essential for the automation and sophistication of signal analysis, particularly in complex environments where real-time processing and accuracy are paramount.

Another dimension to consider is market competition and reach. Post-merger or acquisition, the newly formed entity typically enjoys a larger market footprint and customer base. This expanded reach is vital for the propagation and adoption of innovative signal processing technologies. For example, in telecommunications, firms can leverage an expanded infrastructure to deploy faster and more reliable communication protocols. Through economies of scale, these firms can also offer competitive pricing, making advanced technologies accessible to a broader spectrum of users.

However, this consolidation trend is not without challenges. Integrating distinct corporate cultures and operational frameworks presents a non-trivial challenge. Disparate approaches to project management, employee engagement, and technological paradigms may lead to friction and inefficiencies. Additionally, the monopolistic tendencies resultant from major mergers or acquisitions could potentially stifle competition and innovation in the signal processing industry. Without the competitive pressure, there is a risk of reduced impetus to innovate continually.

An illustrative example is the merger between two leading companies in the semiconductor industry, which has had profound implications on signal processing hardware. By combining their technological expertise and manufacturing capabilities, these firms have been able to reduce costs and expedite the production of high-performance signal processing chips. These chips power an array of devices, from smartphones to sophisticated medical diagnostic equipment, catalyzing advancements across sectors.

Conventional signal processing methodologies benefit significantly from the diversified expertise within merged entities. For instance, in medical imaging, enhanced computational resources and collaborative innovation post-M&A can lead to more accurate diagnostic tools. By integrating superior image processing algorithms with cutting-edge hardware, merged firms can develop systems that provide more precise diagnostics, potentially saving lives through early disease detection.

In summary, mergers and acquisitions in the signal processing industry engender far-reaching consequences, both positive and challenging. The confluence of varied technical proficiencies, expanded market reach, and augmented financial power post-M&A fosters innovation and growth. Nevertheless, the associated challenges, particularly related to cultural integration and potential market monopolies, warrant careful consideration. Properly managed, mergers and acquisitions hold the potential to propel the signal processing industry toward new heights of technological advancement and application diversity.

16.3 Innovation in Corporate Signal Processing Strategies

Corporate signal processing strategies have long been pivotal in transforming raw data into actionable insights, influencing decision-making processes across industries. In recent years, the landscape of signal processing has witnessed a series of innovations that have enabled corporations to enhance their competitive edge. These advancements are chiefly driven by the exponential growth in data generation, heightened computational power, and novel algorithmic techniques.

At the core of modern signal processing innovations is the integration of machine learning algorithms, particularly deep learning. Deep learning's prowess in feature extraction and classification has revolutionized tasks such as speech recognition, image analysis, and predictive maintenance. For instance, convolutional neural networks (CNNs) have substantially improved the accuracy of image and video processing, enabling corporations to achieve high precision in quality control inspections on production lines. By deploying these networks, companies can identify defects in real-time, reducing downtime and enhancing product integrity.

Moreover, the deployment of recurrent neural networks (RNNs) has greatly benefited time-series analysis, a critical component in financial forecasting, inventory management, and customer demand prediction. With architectures like long short-term memory (LSTM), these networks can capture temporal dependencies with high fidelity, facilitating more accurate forecasts and robust decision-making frameworks. This innovation is particularly transformative in sectors such as retail and logistics, where understanding seasonal demand patterns can significantly optimize stock levels and reduce holding costs.

Another stride in signal processing strategies is the increased use of edge computing. Traditionally, data has been processed in centralized cloud environments, but the surge in Internet of Things (IoT) devices has necessitated quicker, more localized processing. Edge computing addresses this need by processing data at the source rather than relying on remote servers. This shift reduces latency, increases data security, and allows for real-time analytics, which is crucial for applications like autonomous vehicles, smart grids, and industrial automation.

Furthermore, advancements in signal processing hardware have complemented these algorithmic innovations. Graphics processing units (GPUs) and application-specific integrated circuits (ASICs), such as Google's Tensor Processing Units (TPUs), have drastically reduced the time required to train complex models. The integration of these powerful processors has accelerated the transition of signal processing solutions from research laboratories to practical, corporate applications.

Signal processing strategies have also benefitted from the increased adoption of quantum computing. Though still nascent, quantum algorithms promise unprecedented computational capabilities, particularly in solving optimization and simulation problems at scales unmanageable by classical computers. For instance, quantum signal processing can potentially enhance cybersecurity measures through more robust encryption techniques, thereby protecting corporate data from increasingly sophisticated cyber threats.

Equally transformative has been the development of robust data fusion techniques, which integrate information from multiple sensors and data sources to provide a more nuanced understanding of the environment. Multimodal data fusion is especially critical in areas like autonomous systems and smart cities, providing a richer, more coherent picture that drives intelligent decision-making. For example, self-driving cars utilize data fusion from lidar, radar, and cameras to create a robust interpretation of surroundings, ensuring safer navigation.

Real-time processing capabilities have also evolved, enabled by innovations in software-defined networking (SDN) and network function virtualization (NFV). These technologies provide the agility and flexibility required to dynamically allocate resources based on demand, ultimately facilitating seamless signal processing in dynamic environments. This capability is particularly crucial in telecommunications, where bandwidth utilization can vary unpredictably.

Moreover, integrating feedback mechanisms into signal processing strategies has become increasingly strategic. By continually learning from outputs and adjusting parameters accordingly, systems can achieve higher accuracy and efficiency over time. This adaptive signal processing is vital in industries with highly variable conditions, such as oil and gas exploration, where real-time data from drilling operations can be used to optimize resource extraction.

In conclusion, the ongoing innovations in signal processing strategies are setting new benchmarks for corporate efficiency and competitiveness. Advances in machine learning, edge computing, specialized hardware, quantum computing, data fusion, real-time processing, and adaptive mechanisms are collectively driving this transformation. These technological strides ensure that corporations can more effectively harness the vast amounts of data at their disposal, turning complex signals into significant opportunities for growth and innovation.

16.4 Corporate Governance in Signal Processing Policy Implementation

Corporate governance holds a pivotal role in the realm of signal processing policy implementation. Effective governance frameworks ensure that signal processing strategies are not only articulated clearly but also executed with an alignment towards the overarching organizational goals. The synthesis of robust corporate governance with signal processing policy implementation fosters transparency, accountability, and strategic coherence.

At the heart of the corporate governance system lies a structure that encompasses various entities such as the board of directors, executive management, and other key stakeholders. These entities are entrusted with key responsibilities that influence the trajectory of signal processing policies within an organization. The board of directors, for instance, takes up a fundamental role by setting the strategic direction and making decisions that align with the organization's long-term objectives. Their insights and decisions help shape the policy frameworks governing signal processing, ensuring that such policies resonate with the company's mission and vision.

Executive management, on the other hand, is charged with the operational facets of implementing these policies. They are involved in translating the board's strategic directives into actionable plans and ensuring that these plans are executed meticulously. This translation often requires a profound understanding of the technical and operational aspects of signal processing, thus mandating collaboration between the technical teams and the governance bodies. As a result, executive management often mediates between the board's overarching goals and the technical team's operational capabilities.

Effective corporate governance in signal processing policy implementation is further characterized by a diligent risk management framework. Given the intricate and often sensitive nature of signal processing, organizations must be adept at identifying, assessing, and mitigating potential risks that may arise. This involves implementing stringent protocols and compliance mechanisms to safeguard against breaches, systemic failures, and external threats. The audit committee, typically a subset of the board, plays a crucial role in this context by ensuring that risk management policies are robust and consistently adhered to. They conduct periodic audits, review compliance reports, and provide feedback to refine and enhance the policy framework.

The importance of stakeholder engagement in corporate governance cannot be overstated. Engaging with various stakeholders, including shareholders, employees, customers, and regulatory bodies, offers myriad benefits. It ensures a broader perspective is considered in policy formulation and implementation, and it promotes greater buy-in and support from those impacted by these policies. Open lines of communication facilitate a flow of information that can lead to more informed decision-making and policy adjustments that reflect stakeholder concerns and inputs.

Transparency forms another pillar within the corporate governance structure critical to signal processing policy implementation. Transparency enables stakeholders to understand the rationale behind certain policies, decisions, and actions. It fosters a culture of trust, as stakeholders feel assured that operations are conducted with integrity and that they can hold management accountable. Public disclosures, detailed reporting, and accessible communication channels are instrumental in maintaining this transparency, particularly in areas as specialized and potentially opaque as signal processing.

Technological advancements and their implications on signal processing necessitate continuous learning and adaptation. Corporate governance structures must, therefore, incorporate mechanisms for regular training and development. This is to ensure that directors, executives, and employees stay abreast of the latest technological trends, regulatory changes, and best practices in signal processing. Continuous education ensures that governance mechanisms remain relevant and effective in navigating the dynamic landscape of signal processing technologies.

Moreover, the alignment of corporate governance practices with ethical standards is essential. Ethical considerations, such as data privacy, equity, and non-discrimination, are critical when implementing signal processing policies, given their potential impact on society. A governance framework imbued with ethical principles ensures that signal processing innovations and implementations contribute positively to societal welfare and are in compliance with legal standards.

In conclusion, the interplay between corporate governance and signal processing policy implementation is a complex yet crucial one. It requires the deliberate orchestration of strategic oversight, risk management, stakeholder engagement, transparency, continuous learning, and ethical adherence. This comprehensive approach not only enhances the efficacy and reliability of signal processing operations but also aligns them with the broader, beneficial goals of the organization and society at large.

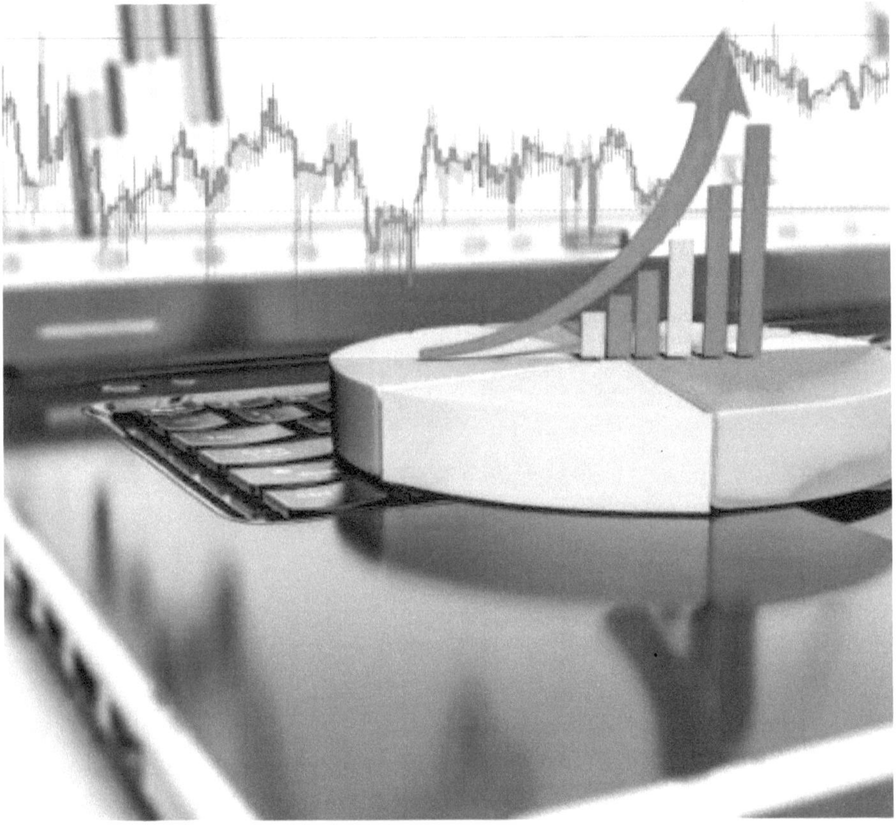

17. Metrics and Evaluation in Signal Processing

The field of signal processing, at its core, revolves around the transformation and analysis of data to extract meaningful information. As technology has rapidly advanced, so too have the methods and tools used in this domain. However, the effectiveness and efficiency of these methods cannot be assumed—they must be scrutinized, validated, and optimized. This is where metrics and evaluation play a pivotal role. By systematically measuring performance through well-defined criteria, we can discern the strengths and limitations of various signal processing techniques. This chapter delves into the essential frameworks and methodologies for evaluating signal processing algorithms, elucidating how performance metrics are developed, interpreted, and applied. Through this exploration, we aim to equip researchers and practitioners with the knowledge necessary to make informed decisions, fostering innovations that are not only theoretically sound but also practically impactful.

17.1 Key Performance Indicators (KPIs) for Signal Processing Success

In the realm of signal processing, Key Performance Indicators (KPIs) are essential metrics that help quantify the success and efficiency of algorithms and systems. These KPIs are particularly vital in ensuring that the designed solutions meet the requirements of specific applications, which can range from communications to biomedical engineering.

One of the primary KPIs in signal processing is the Signal-to-Noise Ratio (SNR). This metric evaluates the level of the desired signal to the background noise present in the system. A higher SNR often indicates a clearer and more reliable signal, which is crucial for applications like speech recognition and audio enhancement. In managing SNR, it is essential to consider both the signal strength and the noise level introduced by the system or the environment.

Another critical KPI is the Bit Error Rate (BER). In digital communication systems, BER measures the number of bit errors divided by the total number of bits transmitted over a given time period. This indicator is fundamental in assessing the integrity and reliability of data transmission. Lower BER values suggest more accurate data transmission, which is especially important in high-stakes fields such as financial transactions and remote sensing.

The third KPI to consider is latency, often critical in real-time signal processing applications. Latency refers to the delay from input to output in a system. In scenarios like live audio streaming or online gaming, high latency can result in poor user experiences or even render the application unusable. Reducing latency involves optimizing algorithm processing times and ensuring efficient system architecture to handle the signal promptly.

Throughput, another essential KPI, measures the amount of data that can be processed by the system within a specific timeframe. Higher throughput signifies that the system can handle larger volumes of data efficiently, which is imperative in applications dealing with high data rates, such as video streaming or large-scale sensor networks.

Accuracy, defined as the closeness of the system's output to the true or expected result, stands as a pivotal KPI in signal processing. Various applications, such as medical imaging or fault detection in engineering systems, rely heavily on the accuracy of the processed signals. Techniques like calibration and advanced algorithmic adjustments are often deployed to enhance accuracy.

Moreover, robustness is a vital KPI for assessing how well a signal processing system can perform under varying and unpredictable conditions. In practice, robustness can refer to the system's ability to maintain performance despite signal degradation, noise, or interference. For instance, in adaptive filtering, the robustness of the algorithm determines its effectiveness in dynamically changing environments.

Energy efficiency is an increasingly important KPI, especially in portable or battery-operated devices. This metric focuses on the amount of power consumed by the signal processing tasks over time. Energy-efficient systems extend the operational life of the device, making this KPI crucial in applications like mobile communication and wearable health monitors.

Scalability, the ability of the system to handle growing amounts of work or its potential to expand to accommodate growth, is another important KPI. In large-scale applications, such as network processing or distributed sensor arrays, scalability ensures that the system can maintain performance levels as the demand or dataset size increases.

Lastly, complexity measures how resource-intensive a particular signal processing algorithm is in terms of computational power and memory usage. Lower complexity is desirable as it typically translates to faster computation and less resource consumption. Efficient algorithms that maintain performance with minimal complexity are particularly valuable in embedded systems and real-time processing applications.

In sum, these KPIs provide a comprehensive framework for evaluating and optimizing signal processing systems. Integrating considerations of SNR, BER, latency, throughput, accuracy, robustness, energy efficiency, scalability, and complexity allows developers to create solutions that are not only technically proficient but also applicable to real-world challenges. As signal processing technologies continue to advance, continuously assessing and refining these KPIs will be crucial in driving innovation and achieving excellence in various applications.

17.2 Measuring the Effectiveness of Signal Processing Strategies

Assessing the effectiveness of signal processing strategies is a critical aspect of ensuring optimal performance in various technological applications, including telecommunications, audio engineering, and biomedical signal analysis. This rigorous evaluation involves a multifaceted approach, employing both quantitative metrics and qualitative assessments to determine how well a particular strategy meets its intended goals.

One fundamental method of evaluation is through quantitative performance metrics. These include Signal-to-Noise Ratio (SNR), Bit Error Rate (BER), Mean Squared Error (MSE), and Total Harmonic Distortion (THD). Each of these metrics provides a numerical indicator of performance that can be compared against benchmarks or competing methods. For instance, a high SNR indicates that the signal is relatively free from noise, affording clearer and more accurate transmission. Similarly, a low BER signifies fewer errors in data transmission, which is crucial for applications that demand high reliability. MSE is another critical metric, particularly in adaptive filtering and predictive modeling, as it quantifies the average of the squares of the errors, providing insight into the fidelity of the signal processing method.

In addition to these metrics, the computational complexity of the signal processing strategy is also a key factor in determining its effectiveness. This involves evaluating factors such as processing speed, memory usage, and power consumption. Modern applications often require real-time processing, which necessitates not only accuracy but also speed. An algorithm that delivers high accuracy but is computationally intensive may not be suitable for real-time applications. Conversely, a faster but less accurate algorithm might be preferable in scenarios where quick decision-making outweighs the need for precision. Therefore, the trade-offs between accuracy and computational complexity must be carefully balanced based on the specific requirements of the application.

Subjective assessments also play a significant role, especially in fields like audio and image processing. These evaluations often involve human judgment to determine the perceived quality of the processed signal. For instance, in audio signal processing, listeners might be asked to rate the clarity, naturalness, and lack of artifacts in processed sound samples. Similarly, in image processing, visual assessments by experts can help gauge the effectiveness of different strategies in maintaining or enhancing image quality. While subjective assessments can introduce variability, they are essential for applications where human perception is the ultimate measure of success.

Experimental validation is another cornerstone of evaluating signal processing strategies. This involves implementing the strategy in real-world scenarios and measuring its performance under actual operating conditions. Laboratory simulations provide valuable insights but may not capture all the nuances and potential challenges encountered in real-world usage. Field tests can uncover issues related to environmental noise, hardware limitations, and other practical considerations that might not be apparent in a controlled setting. The results of such experiments can offer a more comprehensive understanding of the strategy's robustness and reliability.

Moreover, the effectiveness of signal processing strategies can also be assessed through comparative studies. By benchmarking a new strategy against established methods, one can highlight its relative strengths and weaknesses. This comparative approach often involves running multiple strategies on the same dataset or subjecting them to the same conditions and then comparing their performance across various metrics. Such comparisons are invaluable in highlighting advancements and identifying areas for further improvement.

Documentation and transparency are vital components of the evaluation process. Detailed documentation of the methods, assumptions, and conditions under which the evaluations were conducted ensures that the results are reproducible and verifiable. Peer reviews and open-access publications provide additional layers of scrutiny, fostering a collaborative environment where strategies can be continuously refined and optimized based on collective insights.

In conclusion, measuring the effectiveness of signal processing strategies is a comprehensive and dynamic process that leverages a mix of quantitative metrics, qualitative assessments, experimental validations, and comparative analyses. Balancing these different aspects allows for a holistic evaluation, ensuring that the chosen strategy is not only theoretically sound but also practically viable and aligned with user expectations and real-world demands. Advances in this field continue to refine these evaluation techniques, driving the development of more robust and efficient signal processing solutions.

17.3 Benchmarking Signal Processing Performance Across Industries

Benchmarking is a critical tool used to measure the performance of various functions across different industries. In the realm of signal processing, benchmarking involves comparing the efficacy and efficiency of signal processing techniques against established standards or the performance metrics of leading practitioners. Signal processing itself refers to the analysis, interpretation, and manipulation of signals which may include sound, images, and scientific measurements. By benchmarking signal processing performance, industries aim to enhance their capabilities, identify best practices, and implement improvements in their systems.

Signal processing is integral to numerous industries including telecommunications, healthcare, automotive, aerospace, and consumer electronics. Given its wide application, benchmarking across these sectors involves assessing a multitude of parameters such as latency, accuracy, resource consumption, and robustness under various conditions.

In telecommunications, signal processing is crucial for data transmission and reception. The performance can be evaluated by examining metrics such as bit error rate (BER), signal-to-noise ratio (SNR), and latency. For example, the implementation of adaptive filtering techniques can be benchmarked against fixed filtering techniques to determine improvements in noise reduction and signal clarity. By comparing these metrics across different telecommunications providers, best practices can be established, driving the industry toward more reliable and efficient communication systems.

In healthcare, particularly in imaging technologies like MRI and CT scans, signal processing performance directly impacts diagnostic accuracy. Benchmarking here might involve comparing the resolution, processing time, and noise reduction capabilities of different imaging systems. High-performance signal processing can lead to earlier and more accurate diagnoses, enhancing patient care. Companies can benchmark their imaging systems against leading equipment to ensure they provide high-quality images with minimal artifacts or processing delays.

The automotive industry leverages signal processing in systems related to driver assistance and autonomous driving. Performance benchmarks in this sector could include reaction times, accuracy in identifying obstacles, and the ability to operate in diverse environmental conditions. For instance, the object recognition performance of different radar and LIDAR systems can be compared to determine the most reliable solutions for real-time navigation. Benchmarking these systems helps manufacturers integrate the most effective technologies, ensuring safety and efficiency.

Aerospace applications of signal processing encompass radar systems, communication systems, and navigation tools. Critical performance metrics might include detection accuracy, range resolution, and processing speed. Benchmarking can involve simulations and real-world tests to compare different systems under various operational conditions. High-stakes environments like aerospace demand exquisite precision; thus, benchmarking ensures that the most advanced and reliable systems are employed.

Consumer electronics, such as smartphones and home audio systems, rely heavily on signal processing for functions like voice recognition and noise cancellation. Performance metrics relevant to this industry include clarity, response time, and power consumption. Benchmarking might involve comparing the performance of different algorithms or hardware configurations to find the optimal balance between performance and energy efficiency. By consistently benchmarking their products, companies can stay ahead of the competition, offering enhanced features and improving user experience.

Cross-industry benchmarking presents additional challenges due to the differing priorities and operating environments of each sector. However, developing a unified benchmarking framework can provide valuable insights. For example, performance benchmarks in terms of processing speed or accuracy might be universally applicable, with additional industry-specific metrics layered on. Techniques like machine learning are being increasingly adopted across industries; thus, the performance of machine learning-augmented signal processing algorithms can provide a common ground for benchmarking.

Moreover, the exchange of benchmarking results and best practices across industries can drive innovation. Insights gained from the automotive sector's advancements in real-time object recognition could be adapted for use in healthcare for more accurate imaging analysis. Similarly, noise reduction techniques in telecommunications could inspire improved noise cancellation technologies in consumer electronics.

Ultimately, the goal of benchmarking signal processing performance across industries is to foster a competitive environment where continuous improvement is encouraged. By understanding how each industry addresses specific challenges and adopting the best-performing techniques, sectors can substantially enhance their signal processing capabilities, leading to better products, services, and outcomes across the board.

17.4 Auditing and Reviewing Signal Processing Practices

In the realm of digital signal processing (DSP), auditing and reviewing practices are crucial for ensuring the integrity, reliability, and regulatory compliance of signal processing systems. These systems are widely employed in various fields such as telecommunications, audio processing, medical diagnostics, and more. Given the high stakes associated with inaccuracies or failures in these systems, systematic auditing and reviewing processes are indispensable.

Firstly, the foundation of effective auditing in signal processing lies in establishing clear guidelines and standards. Organizations typically follow international standards, such as those from the International Organization for Standardization (ISO) or the Institute of Electrical and Electronics Engineers (IEEE). These standards provide frameworks that include specific requirements for processes, materials, and performance metrics. By adhering to such guidelines, organizations can ensure that their signal processing practices align with industry best practices, thus facilitating easier auditing and review.

The process of auditing signal processing practices usually begins with a comprehensive review of documentation. This documentation includes detailed descriptions of algorithms, development processes, testing protocols, and previous audit reports. Detailed and accurate documentation serves as the backbone for any audit, offering a clear trail of development and modifications that have been made. Auditors analyze this documentation to verify completeness, consistency, and adherence to established protocols. This initial step helps in identifying any obvious discrepancies or areas needing more rigorous examination.

In addition to documentation review, it is vital to conduct code audits. Signal processing often involves complex algorithms written in high-level programming languages such as MATLAB, Python, or C++. An in-depth code audit aims to identify potential issues such as bugs, inefficiencies, or security vulnerabilities. Coding standards and best practices—such as modularity, readability, and appropriate commenting—are scrutinized to ensure maintainability and scalability. Automated tools like static code analyzers can aid in this process, providing preliminary insights that human auditors can further investigate.

Unit testing is another critical component of auditing signal processing practices. This involves testing individual components of the signal processing system in isolation to ensure each performs as expected. Auditors review the scope and comprehensiveness of unit tests, investigating whether they genuinely cover all edge cases and potential failure points. Auditors must verify that tests are regularly updated to account for changes in algorithms or operational requirements.

Another dimension worth investigating is cross-validation and reproducibility. Cross-validation uses different subsets of data to test the robustness of signal processing algorithms, ensuring they perform consistently across varying conditions. Auditors check whether appropriate cross-validation techniques have been employed and whether results are reproducible. Reproducibility ensures that independent runs of the algorithms produce consistent outputs, an essential requirement for scientific rigor and operational reliability.

In practical implementation, signal processing systems often operate under real-time or near-real-time constraints, requiring auditors to take performance metrics into account. Key performance indicators (KPIs) such as latency, throughput, and resource utilization need to be measured and reviewed. Auditors evaluate whether these KPIs meet the predefined standards and operational requirements. Performance bottlenecks identified during the audit can guide subsequent optimization efforts to enhance overall system efficiency.

Security and privacy measures also constitute an essential checkpoint in auditing signal processing practices. As these systems often process sensitive data, auditors must verify that robust encryption, authentication, and access control mechanisms are in place. Vulnerability assessments and penetration tests help identify potential security flaws, ensuring that the system is safeguarded against unauthorized access and data breaches.

Human factors should not be overlooked in the auditing process. It is important to assess the competencies and training levels of the personnel involved in developing and maintaining the signal processing systems.

Adequate training ensures that individuals are well-versed in both theoretical and practical aspects, including standard operating procedures (SOPs), safety protocols, and troubleshooting techniques.

To ensure the auditing process is objective and unbiased, it is often beneficial to engage third-party auditors. External auditors bring a fresh perspective and may identify issues overlooked by internal teams. Their independent assessment adds credibility to the audit results and increases confidence among stakeholders.

In summary, auditing and reviewing signal processing practices are multi-faceted endeavors that require meticulous attention to detail across documentation, coding standards, unit testing, cross-validation, performance metrics, security, and human factors. Effective auditing ensures that signal processing systems are reliable, efficient, and secure, meeting both operational needs and regulatory requirements.

18. Barriers to Effective Signal Processing

Signal processing stands as a cornerstone of modern technology, underlying the functionality of everything from mobile communication systems to sophisticated medical imaging techniques. However, the path to robust and reliable signal processing is fraught with challenges. These impediments can stem from various sources, including environmental factors, hardware limitations, and algorithmic complexities. Understanding these barriers is critical for both researchers and practitioners striving to enhance the efficacy of signal processing systems.

This chapter delves into the multifaceted obstacles that hinder effective signal processing. We will explore the impact of noise, interference, and distortion on signal integrity, as well as examine the limitations imposed by hardware constraints and computational resources. Additionally, we will address the complexities inherent in adaptive algorithms and real-time processing requirements. Through a comprehensive examination of these impediments, we aim to equip readers with a deeper understanding of the current landscape and inspire innovative solutions to overcome these persistent barriers.

18.1 Common Challenges in Managing Signal Processing

Signal processing, pivotal in various applications across multiple fields such as communications, audio engineering, and medical imaging, is fraught with numerous challenges that can make efficient and accurate management a complex undertaking. Below are some common issues encountered in managing signal processing tasks.

Complexity of Signals

One of the foremost challenges in signal processing is the inherent complexity of the signals themselves. Real-world signals often contain a multitude of frequencies and varying amplitudes, making their analysis and interpretation a difficult task. For example, signals captured from natural scenes or biomedical sources comprise noise, distortions, and artifacts that require sophisticated algorithms for proper processing and filtering.

Noise Reduction

Noise, the unwanted random addition to a signal, poses a significant problem. Effective signal processing mandates the reduction or elimination of noise without compromising the integrity of the original signal. This often requires advanced techniques such as the use of adaptive filters, which adjust their characteristics in real-time based on the signal and noise properties, or employing machine learning algorithms trained to distinguish between relevant signals and noise.

Real-Time Processing

Another crucial challenge is real-time processing, particularly relevant in applications such as speech recognition, autonomous vehicles, and live broadcast systems. Processing signals in real time requires highly efficient algorithms capable of executing within stringent time constraints. Delays introduced by signal processing can render a system useless or, in the case of safety-critical applications, hazardous.

Resource Constraints

Managing resource constraints—such as limited computational power, memory, and battery life—is a common issue, especially in portable or embedded systems. Algorithms must be optimized to run efficiently on hardware with limited capabilities, sometimes demanding creative approaches such as downsampling signals to reduce the amount of data that needs to be processed or developing lightweight algorithms that balance accuracy with computational efficiency.

High Dimensionality

Signals can be high-dimensional, presenting challenges in terms of both storage and processing. For instance, image and video signals involve vast amounts of data that must be processed rapidly for applications such as real-time surveillance. Techniques like dimensionality reduction through Principal Component Analysis (PCA) or the use of convolutional neural networks (CNNs) are essential but computationally intensive, requiring effective management to avoid overload.

Algorithm Robustness

The robustness of signal processing algorithms is also critical. Algorithms need to be resilient to variations and distortions in the input signals. This includes addressing issues such as changes in signal amplitude, frequency variations, and phase shifts. Ensuring robustness often involves extensive testing under various scenarios and incorporating mechanisms to adapt to unpredictable changes.

Synchronization Issues

In applications involving multiple signal sources or channels, synchronization is key. Desynchronization can lead to inaccurate data interpretation and poor performance of the overall system. Synchronizing multiple signals, such as audio and video streams or sensor data from a distributed network, requires precise timing control and sophisticated algorithms to align the data correctly.

Validation and Testing

Adequate validation and testing of signal processing systems are crucial yet challenging. Simulating real-world conditions accurately to validate the performance of signal processing algorithms is often difficult. Real-time testing environments might be necessary, which are resource-intensive and complex to set up. Moreover, the variability of real-world signals means extensive testing across a wide range of scenarios to ensure reliability and performance.

Scalability

Scalability is another common challenge, especially in systems expected to handle increasing amounts of data over time. Scalable algorithms need to maintain efficiency and accuracy as the volume of data grows. This is particularly significant in big data applications where processing

Interference and Crosstalk

In environments with multiple signal sources, interference and crosstalk between channels are prevalent issues. Effective signal processing must distinguish between overlapping signals and isolate the desired signal without mutual interference. Techniques such as MIMO (Multiple Input Multiple Output) are employed in wireless communication to handle such interference, though they add another layer of complexity to the processing tasks.

Evolving Standards and Technologies

Lastly, the rapid evolution of standards and technologies in fields like telecommunications necessitates continual updates to signal processing systems. Keeping up with new protocols, modulation schemes, and signal processing techniques demands a flexible and modular approach to system design, allowing updates and changes to be incorporated with minimal disruption.

Addressing these challenges involves a multidisciplinary approach blending expertise in engineering, mathematics, and computer science, ensuring that signal processing systems meet the high standards demanded by contemporary applications.

18.2 Overcoming Resistance to Signal Processing Initiatives

Addressing resistance to signal processing initiatives in any organization can be seen as an exercise in change management, where understanding human and organizational behavior is crucial. Resistance often stems from various factors, such as fear of the unknown, lack of understanding of the benefits, concerns about job security, and comfort with the status quo. Overcoming this resistance effectively requires a strategic approach, incorporating clear communication, stakeholder engagement, and consistent support structures.

Before delving into the specifics of overcoming resistance, it's essential to understand the nuanced reasons behind it. Organizational inertia—remaining in a current state due to the energy required to change—can make significant technological shifts challenging. Employees who have grown accustomed to particular workflows may fear that new signal processing initiatives might disrupt their routines or render their skills obsolete. Addressing these concerns requires empathy and a methodical approach to change management.

One fundamental strategy to mitigate resistance is fostering a culture of communication. Information dissemination about the new initiatives should be transparent and comprehensive. This involves clearly articulating the benefits of the new signal processing methods, demonstrating how these changes can lead to better performance outcomes, improved efficiency, and overall organizational growth. When employees understand the 'why' behind the change—how it aligns with the organization's goals and their roles—they are more likely to buy into the initiative.

Training and education play pivotal roles in easing resistance. Providing ample training sessions and resources to support the transition can alleviate fears associated with skill gaps. When employees are confident in their ability to use new signal processing tools and technologies, their anxiety reduces, leading to a more favorable reception. Moreover, involving employees in pilot programs can be an effective method to demonstrate the efficacy and user-friendliness of new systems. Hands-on experience can often dispel myths and misconceptions faster than theoretical explanations.

Engaging key stakeholders early in the process is also an essential component. Often, resistance can be preemptively reduced when influential members within the organization advocate for the change. These stakeholders, who are respected and trusted by their peers, can act as champions of the initiative, providing credibility and reducing skepticism. Establishing a coalition of support across various departments can help create a unified front, making the initiative appear as a collective step forward rather than a top-down imposition.

Another critical factor in managing resistance is active listening. Providing forums—such as town hall meetings, feedback sessions, and suggestion boxes—enables employees to voice their concerns and suggestions. This feedback should not merely be collected but also visibly acted upon. When employees see that their input has a tangible impact on the initiative's implementation, trust is built, and resistance diminishes. This iterative process of feedback and adaptation also ensures that the signal processing initiative is continuously refined to meet the needs of its users, thereby increasing its effectiveness and acceptance.

Incentivizing change can also catalyze acceptance. Offering rewards, recognition, or even career advancement opportunities tied to the adoption of new signal processing tools can motivate employees. However, it is essential that these incentives are perceived as fair and attainable, as misaligned or overly competitive incentive structures can lead to adverse effects, such as unhealthy competition or demotivation among non-recipients.

Communication should continue even after the initial roll-out. Regular updates on progress, success stories, and ongoing challenges keep the workforce informed and connected to the initiative's journey. This sustained communication helps to maintain momentum and addresses any emerging resistance along the way.

In conclusion, overcoming resistance to signal processing initiatives involves a multifaceted approach rooted in transparency, education, stakeholder engagement, active listening, and continuous support. Understanding the underlying fears and concerns, and addressing them systematically, can transform resistance into acceptance and enthusiasm. By creating an environment where employees feel informed, involved, and valued, organizations can ensure that their signal processing initiatives not only survive but thrive.

18.3 Regulatory and Compliance Challenges in Signal Processing

The realm of signal processing, which encompasses the manipulation and analysis of signals to extract valuable information or to enhance transmission and storage, presents a multitude of regulatory and compliance challenges. These challenges arise from specific normative requirements related to privacy, security, intellectual property, and industry-specific regulations, among others. As the technology evolves, so do the complexities and implications of compliance in this field.

Privacy is one of the most significant regulatory challenges in signal processing. This issue is particularly pronounced in applications involving the processing of personal data, such as in telecommunications, healthcare, and surveillance. Regulations like the General Data Protection Regulation (GDPR) in Europe and the California Consumer Privacy Act (CCPA) in the United States impose stringent requirements on how personal data is to be handled. Compliance entails ensuring that data is anonymized or pseudonymized when necessary, that individuals' consent is obtained for processing their data, and that there is transparency about how their data is used. Failure to conform to these regulations can result in heavy fines and significant reputational damage for organizations.

In industries such as healthcare, data integrity and security are paramount. The Health Insurance Portability and Accountability Act (HIPAA) in the United States sets the standards for protecting sensitive patient data. Signal processing applications in this field must employ robust encryption methods to secure data during transmission and storage. Moreover, there must be comprehensive audit logs and access control mechanisms to monitor and restrict access to data. The integration of these security measures in signal processing algorithms and systems can be technically challenging but is essential for compliance.

Intellectual property rights also introduce compliance challenges in signal processing. Algorithms and methods used in this field may be patented, and developers must ensure that their work does not infringe on existing patents. This requires thorough research and, often, the acquisition of licenses to utilize particular technologies. Additionally, open-source software components used in signal processing applications may come with their own licensing requirements, which must be carefully adhered to avoid legal repercussions. Organizations need to maintain detailed records and ensure their developers are educated about these requirements to avert potential infringements.

The financial industry, where signal processing is used for predictive analytics and automated trading, is subject to regulations that demand transparency and accountability. Regulatory bodies such as the Securities and Exchange Commission (SEC) in the United States mandate that financial institutions demonstrate the integrity of their data and the validity of the models they employ. This necessitates thorough documentation of the signal processing techniques used and the establishment of mechanisms to audit and verify algorithmic decisions. Furthermore, measures must be taken to mitigate risks associated with algorithmic trading, such as implementing circuit breakers to prevent market crashes.

In addition to these domain-specific regulations, general cybersecurity frameworks such as the National Institute of Standards and Technology (NIST) Cybersecurity Framework or the International Organization for Standardization (ISO) standards outline best practices that must be followed to safeguard systems that utilize signal processing. These frameworks often require a holistic approach to security, encompassing network security, physical security, and the secure development lifecycle of software products.

Compliance in signal processing is not a one-time effort but an ongoing process. As new regulations emerge and existing ones evolve, organizations must continuously monitor and update their practices. This often involves appointing dedicated compliance officers or teams to stay abreast of regulatory changes, conducting regular training for staff, and performing periodic audits to ensure that all aspects of the signal processing system adhere to the relevant regulations.

Furthermore, as the field of signal processing incorporates emerging technologies like artificial intelligence and machine learning, new ethical and regulatory considerations come into play. There are growing concerns about algorithmic bias and fairness, and regulators are starting to scrutinize the use of these technologies more closely. Signal processing professionals must be prepared to demonstrate that their systems are not only effective but also ethically sound and compliant with emerging standards.

In conclusion, navigating the regulatory and compliance landscape in signal processing is a complex and multifaceted endeavor. Organizations must be vigilant in understanding the specific requirements of their industry, adopting best practices for privacy and security, respecting intellectual property laws, and continuously updating their practices in response to new regulations and technologies. By doing so, they can not only avoid legal pitfalls but also build trust with their clients and users, fostering an environment where technological advancements can thrive responsibly.

18.4 Strategies for Overcoming Signal Processing Barriers

Signal processing stands as the cornerstone of modern digital communications, multimedia systems, control systems, and countless other technological arenas. However, various barriers often impede its efficacy, raising challenges that necessitate novel solutions for sustained progress. The strategies for overcoming these signal processing barriers encompass a mix of advanced algorithms, innovative hardware designs, and sophisticated software techniques.

To commence, one fundamental tactic involves the embrace of adaptive filtering techniques. Adaptive filters are dynamic entities capable of adjusting their parameters in real-time to optimize performance in varying environments. This approach is particularly pertinent in dealing with noise, interference, and signal distortion. For example, Least Mean Squares (LMS) and Recursive Least Squares (RLS) algorithms are prevalent adaptive methods that modify filter coefficients to minimize the error signal, thus enhancing signal clarity and fidelity.

Secondly, leveraging machine learning and artificial intelligence (AI) has become a formidable strategy. Machine learning algorithms, particularly those involving neural networks, have proven their strength in pattern recognition, noise reduction, and predictive maintenance of signal processing systems. Convolutional Neural Networks (CNNs) and Recurrent Neural Networks (RNNs) can autonomously learn patterns within vast datasets, thus optimizing signal detection, classification, and interpretation tasks. These AI-driven methodologies offer a shift from traditional rule-based approaches to more sophisticated, data-driven models capable of handling complex, non-linear relationships inherent in signal processing.

Furthermore, advanced compression techniques are imperative for overcoming the barrier of limited bandwidth. Efficient data compression methods like Discrete Cosine Transform (DCT), Wavelet Transform, and more recently, Deep Learning-based compression algorithms have revolutionized the way digital signals are transmitted and stored. By significantly reducing the size of data without compromising on quality, these techniques facilitate quicker transmission rates and save storage space, a crucial advantage in fields such as online streaming and cloud computing.

Moreover, enhancing signal quality through modulation and coding strategies is another vital approach. Techniques such as Quadrature Amplitude Modulation (QAM) and Orthogonal Frequency Division Multiplexing (OFDM) optimize the use of available bandwidth, improving the robustness of signal transmission, particularly in wireless communications. Forward Error Correction (FEC) codes such as Turbo codes and Low-Density Parity-Check (LDPC) codes also play a critical role in detecting and correcting errors in received signals, thereby maintaining the integrity and reliability of the communication system.

The implementation of robust cybersecurity measures for signal processing systems is equally significant. As signal processing applications increasingly interface with the internet and other networks, they become susceptible to cyber threats. Employing robust encryption algorithms, secure communication protocols, and intrusion detection systems can safeguard signal processing operations from unauthorized access, data breaches, and other malicious activities.

In the hardware domain, the development of more potent and energy-efficient processors is crucial. Digital Signal Processors (DSPs) and Field-Programmable Gate Arrays (FPGAs) are being continually refined to handle highly complex computations at rapid speeds while minimizing power consumption. Emerging technologies such as quantum computing also hold promise for exponentially enhancing signal processing capabilities far beyond the limitations of classical computing systems.

Real-time signal processing is another demanding area, necessitating precise timing and synchronization. Implementing parallel processing architectures and exploiting multi-core processors can manage the extensive computational requirements by distributing tasks efficiently. Additionally, advancements in software frameworks and real-time operating systems specifically optimized for signal processing tasks contribute to meeting stringent real-time performance goals.

Lastly, interdisciplinary collaboration is a strategy that cannot be overstated. Signal processing benefits immensely from cross-pollination with fields such as mathematics, physics, computer science, and engineering. This collaborative approach fosters the development of holistic solutions that address diverse aspects of signal processing barriers.

In summary, the synergy of adaptive filtering, artificial intelligence, advanced compression techniques, modulation and coding strategies, stringent cybersecurity measures, sophisticated hardware development, real-time processing capabilities, and interdisciplinary collaboration collectively form a powerful arsenal. These strategies not only surmount current signal processing barriers but also pave the way for future innovations, thereby expanding the horizons of what can be achieved in this pivotal domain.

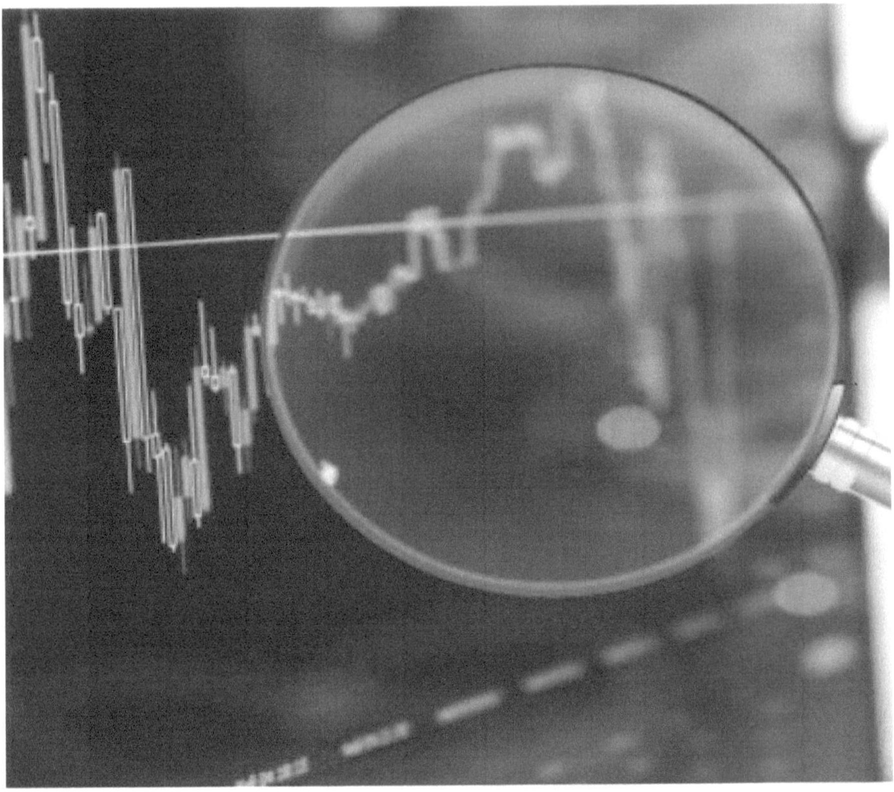

19. Signal Processing and Public Policy

In our increasingly interconnected world, the interaction between technological advancements and regulatory frameworks is both inevitable and essential. This chapter delves into a nuanced examination of signal processing, a cornerstone of modern communication, and how it intersects with public policy. As technology evolves at a rapid pace, the legal and regulatory landscape must adapt to address new challenges and opportunities. From privacy concerns and data security to the implications of digital surveillance and spectrum management, the dialogue between engineers and policymakers has never been more critical. By exploring these interwoven domains, we aim to illuminate the complexities and encourage informed decision-making in the pursuit of societal progress and technological integrity.

19.1 The Role of Government in Shaping Signal Processing Regulations

The role of government in shaping signal processing regulations is multifaceted and vital for the coherent functioning of telecommunications, cybersecurity, and data privacy landscapes. Signal processing, the technique of analyzing, modifying, and synthesizing signals such as audio, video, or sensor data, is integral to numerous sectors including telecommunications, healthcare, automotive industries, and entertainment. Due to its wide-reaching applications and potential for abuse, government intervention becomes necessary to ensure the proper, ethical, and safe use of signal processing technologies.

One of the primary roles of government in this regard is to establish a regulatory framework that dictates the permissible uses and activities concerning signal processing technologies. By formulating and enforcing legislation, governments can set standards that ensure signal processing systems are developed and utilized in ways that respect user privacy, ensure fair competition, and protect national security. For example, laws such as the General Data Protection Regulation (GDPR) in the European Union provide a legal framework that imposes requirements on how signal data must be processed, stored, and shared, thereby safeguarding individual data rights.

Furthermore, the government also plays an instrumental role in standardizing technical protocols. Regulatory authorities like the Federal Communications Commission (FCC) in the United States or the European Telecommunications Standards Institute (ETSI) establish technical standards that equipment manufacturers, telecommunications providers, and associated entities must adhere to. These standards ensure interoperability between different systems and devices, which is crucial for maintaining robust communication networks and enabling seamless data transmission.

Another significant aspect of government involvement is the allocation and management of the electromagnetic spectrum, an essential resource for wireless signal transmission. Governments hold the authority to regulate spectrum allocation, ensuring that it is used efficiently and equitably among various stakeholders such as military, public safety services, commercial entities, and private sectors. Auctions for spectrum licenses, permits, and adherence to specific frequency bands are overseen by government entities to prevent interference and optimize resource utilization.

Cybersecurity is another critical domain where government regulation is necessary. In an era where signal processing underpins numerous cyber-physical systems, safeguarding these systems from unauthorized access and malicious attacks is of paramount importance. Governments establish cybersecurity policies and guidelines that mandate robust encryption standards and secure data transmission practices, thus shielding critical infrastructures from vulnerabilities.

Moreover, government intervention is also essential in promoting innovation and research in signal processing technologies. Public funding for research and development through grants, tax incentives, and collaborative programs between governmental research institutions and private entities is pivotal for fostering technological advancements. Governments may also institute intellectual property laws that protect innovations in signal processing, ensuring that inventors and developers can secure exclusive rights to their creations.

In settings where signal processing technologies intersect with ethical and societal considerations, such as in surveillance or biometric identification, the government's role extends to protecting civil liberties. Regulatory measures are necessary to ensure that these technologies are not misused in ways that infringe upon individual freedoms or lead to discriminatory practices. For instance, regulations may mandate transparency, accountability, and oversight mechanisms to prevent the misuse of surveillance technologies.

In summary, the government's role in shaping signal processing regulations is comprehensive and dynamic, encompassing the development of legal frameworks, establishment of technical standards, management of the electromagnetic spectrum, fortification of cybersecurity, promotion of innovation, and protection of civil liberties. By executing these roles effectively, governments can ensure that signal processing technologies contribute positively to societal progress while mitigating associated risks and challenges.

19.2 Public-Private Partnerships in Signal Processing Management

Public-private partnerships (PPPs) in signal processing management represent an evolving and synergistic approach to addressing some of the most pressing challenges in technology and infrastructure today. This collaborative model leverages the strengths of both the public and private sectors to advance the development, deployment, and management of signal processing technologies. These partnerships are particularly relevant in fields such as telecommunications, digital signal processing for healthcare, intelligent transportation systems, and defense applications, where the integration of cutting-edge technology and robust infrastructure is critical.

Signal processing encompasses a range of activities that involve the analysis, interpretation, and manipulation of signals to improve their efficiency, reliability, and quality. This technology finds applications in numerous domains, making it a focal point for innovation and investment. The increasing complexity and the need for high precision in signal processing tasks necessitate substantial investment in research and development, robust infrastructure, and skilled human resources. This is where PPPs come into play, combining public sector oversight and funding with private sector innovation and expertise to create a powerful partnership.

One of the primary advantages of PPPs in signal processing management is the pooling of resources and expertise. The public sector typically provides funding, regulatory support, and long-term strategic planning. Government agencies may also facilitate access to critical infrastructure and national assets, such as spectral bandwidth and data from public institutions. On the other hand, the private sector brings agility, specialized knowledge, and the ability to innovate rapidly. Private companies are often at the forefront of technological advancements and can introduce state-of-the-art solutions more swiftly than the public sector alone.

The collaborative nature of PPPs also helps mitigate the risks associated with large-scale signal processing projects. These projects often require significant upfront investment and have long development cycles, making them financially risky. By partnering with the private sector, the financial burden is shared, and risks are distributed. Additionally, private sector involvement often leads to a more disciplined approach to project management, as private entities are accountable to shareholders and investors, driving efficiency and effectiveness.

In the context of telecommunications, PPPs have been instrumental in expanding broadband connectivity and enhancing digital communication networks. Governments worldwide recognize the critical importance of advanced communication infrastructure for economic development and social well-being. Through PPPs, substantial investments are channeled into developing robust, high-speed internet networks, extending connectivity to underserved and rural areas, and upgrading legacy systems to support new technologies like 5G. These partnerships not only help bridge the digital divide but also stimulate economic growth by fostering innovation, enabling new business models, and creating job opportunities.

In healthcare, signal processing technologies are pivotal in medical imaging, patient monitoring, and diagnostic systems. PPPs in this sector have led to advancements in telemedicine, wearable health devices, and data analytics, revolutionizing patient care and health management. Public health systems benefit from private sector innovation and expertise in developing sophisticated medical technologies, thereby improving patient outcomes and optimizing operational efficiencies.

Intelligent transportation systems (ITS) are another domain where PPPs play a crucial role. The integration of signal processing technologies in traffic management, autonomous vehicles, and smart infrastructure requires substantial investment and multidisciplinary expertise. By collaborating, public agencies and private companies can design and implement systems that enhance traffic flow, reduce accidents, and optimize transportation logistics. These systems rely heavily on accurate, real-time signal processing for data from various sources, including sensors, cameras, and communication networks.

In defense applications, signal processing is vital for surveillance, reconnaissance, communication, and electronic warfare. PPPs in the defense sector often involve collaborations between government defense departments and private defense contractors. These partnerships facilitate the development of advanced radar systems, encrypted communication technologies, and signal intelligence tools. The private sector's capacity for rapid innovation and deployment is particularly valuable in keeping pace with evolving security threats.

In conclusion, public-private partnerships in signal processing management offer a strategic framework that enhances the development and implementation of critical technologies. By leveraging the complementary strengths of the public and private sectors, PPPs foster innovation, distribute risks, and ensure the efficient execution of projects, ultimately leading to technological advancements and improved societal outcomes.

19.3 Navigating Regulatory Changes in Signal Processing

Navigating regulatory changes in the field of signal processing can be a complex and multifaceted challenge. With the rapid pace of technological advancements and the increasing influence of various regulatory bodies, professionals in this industry must remain vigilant and adaptable. This section delves into the intricacies of understanding, adapting to, and capitalizing on regulatory changes within the signal processing domain.

Signal processing, fundamentally, involves the analysis, modification, and synthesis of signals such as sound, images, and scientific measurements. Over the years, its applications have expanded into diverse fields, including telecommunications, healthcare, and defense. Consequently, the landscape of regulations governing these applications has become equally diverse and stringent.

One primary challenge arises from the global nature of signal processing applications. Different countries and regions have their distinct regulatory frameworks, each with its unique set of rules and standards. For instance, the United States Federal Communications Commission (FCC) sets different standards compared with the European Telecommunications Standards Institute (ETSI) in Europe. Compliance with these varied standards requires a deep understanding of local regulations and a strategic approach to ensure products and solutions meet all necessary criteria.

Moreover, these regulations are not static. They evolve in response to technological advancements, emerging threats, and changing societal needs. Cybersecurity, for instance, has become a paramount concern in signal processing, particularly in the context of data transmission and storage. Regulatory bodies continuously update their guidelines to address these concerns. The General Data Protection Regulation (GDPR) in Europe, which has strict requirements for data protection and privacy, exemplifies such developments. Compliance with GDPR necessitates more robust encryption and secure signal processing methods, thereby impacting the design and implementation of signal processing systems.

In the healthcare sector, the integration of signal processing technologies into medical devices introduces another layer of regulatory complexity. Devices such as MRI machines, ECG monitors, and wearable health trackers rely heavily on signal processing algorithms. Regulatory bodies like the U.S. Food and Drug Administration (FDA) and the European Medicines Agency (EMA) impose rigorous testing and validation requirements to ensure these technologies are safe and effective. Keeping abreast of these regulations and incorporating compliance into the development lifecycle can be a significant yet essential endeavor for companies operating in this space.

Adapting to these changes often demands a multifaceted approach. Firstly, continuous education and training are crucial. Professionals in the field must stay updated with the latest regulatory trends and standards through workshops, seminars, and courses. Many organizations also invest in compliance teams or consultants specializing in regulatory affairs to navigate this complex landscape efficiently.

Secondly, collaboration and communication within the industry play a pivotal role. Companies can benefit from engaging with regulatory authorities during the early stages of product development. This proactive approach can provide valuable insights into compliance expectations and allow for adjustments before significant investments are made. Additionally, participating in industry consortia and standardization committees enables stakeholders to stay at the forefront of regulatory changes and contribute to shaping future regulations.

Lastly, leveraging technology itself can be an effective strategy for managing regulatory changes. Advanced tools and software solutions designed for compliance management can automate and streamline various aspects of regulatory adherence. These tools can track changes in regulations, assist in documenting compliance efforts, and generate necessary reports for regulatory submissions.

In conclusion, the landscape of signal processing is profoundly influenced by an ever-evolving regulatory environment. Understanding and navigating these regulatory changes require a comprehensive approach involving continuous education, proactive engagement with regulatory bodies, industry collaboration, and the strategic use of technology. Adapting to these changes not only ensures compliance but also fosters innovation and leadership in the rapidly advancing world of signal processing.

19.4 Challenges in Aligning Signal Processing with Public Policy

Aligning signal processing with public policy presents a multifaceted set of challenges that span technological, ethical, legal, and societal considerations. Signal processing, the analysis, interpretation, and manipulation of signals, is pervasive in modern technologies, from telecommunications to healthcare. While its advancements offer significant benefits, integrating these technological strides within the framework of public policy is complex and intricate. Understanding these challenges is imperative for policymakers, technologists, and society as a whole.

One of the primary challenges is the rapid pace of technological advancement in signal processing compared to the slower, more deliberative process of public policy formulation. Technological innovations move at breakneck speeds, driven by market forces and competitive pressures. Public policy, on the other hand, typically involves extensive debate, stakeholder engagement, and legislative procedures. This discrepancy can result in regulations that are outdated by the time they are implemented, failing to address the current capabilities and implications of signal processing technologies.

Furthermore, the technical complexity of signal processing creates a considerable knowledge gap between technologists and policymakers. Signal processing involves sophisticated algorithms, data analysis techniques, and mathematical models that are not easily understood without specialized training. This gap can lead to misunderstandings and miscommunications during the policy formulation process. Policymakers may struggle to grasp the full implications of technological details, while technologists might lack insights into the broader socio-legal context, resulting in policies that are either overly restrictive or insufficiently protective.

Another significant challenge is the ethical dimension of signal processing applications. Technologies such as facial recognition, automated decision systems, and surveillance heavily rely on advanced signal processing techniques. These applications raise serious ethical concerns related to privacy, consent, and potential biases. For instance, facial recognition systems have demonstrated biases against certain demographic groups, leading to unintentional discrimination. Crafting public policies that mitigate these ethical concerns while still leveraging the benefits of signal processing technologies is a daunting task. It requires a balanced approach that considers the rights and liberties of individuals while addressing the societal benefits.

The global nature of technology further complicates alignment with public policy. Signal processing technologies do not adhere to geographic boundaries—they are developed, deployed, and utilized across the globe. This international aspect poses significant regulatory challenges. Different countries have varying legal frameworks, cultural norms, and standards for technology use. Harmonizing signal processing policies across jurisdictions is a complex endeavor requiring international cooperation and standardization. Policymakers must navigate between protecting national interests and complying with international agreements.

Privacy and data protection constitute another critical arena for aligning signal processing with public policy. Signal processing often involves the collection and analysis of vast amounts of data, some of which can be highly personal or sensitive. Ensuring that data governance policies keep pace with technological capabilities is vital. Public policies must address issues such as informed consent, data ownership, the right to be forgotten, and data security. The challenge lies in creating robust legal frameworks that safeguard individual privacy without stifacing technological innovation.

Economic considerations also pose a challenge in this alignment. Governments must balance fostering innovation and competitiveness in the tech industry with ensuring societal wellbeing and ethical standards. Over-regulation can stymie technological progress and economic growth but under-regulation can lead to misuse and harm. Striking this balance requires a nuanced understanding of both economic and technological landscapes, necessitating collaboration between economists, technologists, and policymakers.

Lastly, the cultural implications of signal processing technologies should not be overlooked. Social acceptance and trust in these technologies are crucial for their integration into everyday life. Public ignorance or skepticism can hinder the progress and adoption of beneficial technologies. Public policies must therefore incorporate strategies for public education and engagement, promoting awareness of the benefits and risks associated with signal processing technologies.

In conclusion, the alignment of signal processing with public policy involves navigating a myriad of challenges, from the technical and ethical to the economic and cultural. It requires a multidisciplinary approach, integrating insights from technologists, ethicists, economists, and public policy experts. Only through such comprehensive collaboration can society harness the full potential of signal processing technologies while safeguarding public interest and ethical standards.

230

20. Human Capital in Signal Processing

In the ever-evolving landscape of engineering and technology, the role of human capital remains a cornerstone of innovation and progress. Signal processing, a multifaceted field that encompasses diverse applications such as communications, audio, image and video processing, and biomedical engineering, is no exception to this axiom. The impact of human expertise, creativity, and teamwork is paramount in driving advancements and solving complex challenges within the discipline.

This chapter delves into the pivotal contributions of human capital in the realm of signal processing. It examines the interplay between education, skill acquisition, organizational dynamics, and the continuous learning required to stay abreast of technological advancements. By exploring case studies and real-world examples, we aim to highlight the ways in which individuals and teams have pushed the boundaries of what is possible, transforming theoretical concepts into practical, impactful solutions.

Moreover, this chapter will address the critical importance of interdisciplinary collaboration, emphasizing how the synergy between different fields and the integration of diverse perspectives can lead to groundbreaking innovations. From the nurturing of talent in academic institutions to the cultivation of expertise within industry settings, we will uncover the essential elements that contribute to the thriving ecosystem of human capital in signal processing.

As we unravel these themes, readers will gain a comprehensive understanding of the human factors that drive success in this dynamic field. Through the lens of human capital, we seek to inspire and inform, providing a roadmap for both current and aspiring professionals who aim to make meaningful contributions to the world of signal processing.

20.1 Building Expertise in Signal Processing Management

Building expertise in signal processing management requires an intricate blend of technical knowledge, practical experience, and continuous learning. Signal processing itself is a multifaceted field, encompassing the analysis, modification, and synthesis of signals such as sound, electromagnetic waves, images, and sensor data. Effective management of these processes necessitates not only a fundamental understanding of the underlying theories but also proficiency in deploying industry-standard tools and methodologies.

To develop expertise in this domain, professionals must start with a strong educational foundation. Formal education in electrical engineering, computer science, or a related field is typically the bedrock upon which signal processing knowledge is built. Core coursework often includes subjects such as digital signal processing (DSP), linear algebra, probability, and statistics. Mastery of these subjects provides the necessary theoretical background to understand complex signal processing concepts.

Practical experience is equally important in cultivating expertise. Hands-on experience with signal processing applications, such as speech recognition, image compression, and communications systems, allows individuals to apply theoretical knowledge to real-world problems. Many professionals gain this experience through internships, co-op programs, or laboratory work during their undergraduate and graduate studies. Additionally, involvement in research projects can offer deep insights into cutting-edge technologies and methodologies.

Advanced learning and specialization often follow initial education and experience. Graduate programs focusing on signal processing offer more in-depth study, with courses covering advanced topics such as adaptive filtering, time-frequency analysis, and machine learning for signal processing. These programs not only expand technical knowledge but also foster critical thinking and problem-solving skills.

Staying current with industry advancements is crucial, given the rapid pace of technological change in signal processing. Continuous professional development through workshops, seminars, and conferences allows practitioners to stay informed about the latest research findings, tools, and best practices. Furthermore, publishing research in peer-reviewed journals and presenting at conferences can establish professionals as thought leaders in the field.

A variety of software tools and programming languages are integral to signal processing management. Competence in MATLAB, Python, R, and other relevant software is essential for designing and implementing signal processing algorithms. These tools offer powerful libraries and frameworks that simplify complex computations and signal manipulations. Moreover, familiarity with embedded systems and hardware descriptions languages (HDLs) such as VHDL or Verilog is crucial for those working on hardware implementations of signal processing algorithms.

Understanding industry standards and regulatory requirements is also important. Signal processing applications often need to comply with specific standards, particularly in fields like telecommunications, medical devices, and aerospace. Knowledge of these standards ensures that solutions are not only effective but also compliant with legal and quality benchmarks.

Effective management of signal processing projects requires strong organizational and leadership skills. Project management methodologies such as Agile, Scrum, or Six Sigma can be beneficial in managing timelines, resources, and stakeholder expectations. Soft skills like communication, teamwork, and problem-solving are equally important, as they facilitate collaboration and effective decision-making.

Mentorship and networking play vital roles in professional growth. Experienced mentors can provide guidance, share insights, and offer feedback based on their own experiences. Networking with peers through professional organizations, online forums, and social media can create opportunities for collaboration and knowledge exchange.

Finally, a proactive approach to learning and adaptability is essential. As new technologies emerge and existing technologies evolve, continuing to refine skills and knowledge is imperative. Learning new programming languages, exploring innovative algorithms, and experimenting with emerging hardware are ways to stay at the forefront of the field.

In summary, building expertise in signal processing management involves a combination of comprehensive education, practical experience, continuous learning, and professional development. Proficiency in both theoretical concepts and practical applications, together with the ability to stay current with industry trends, is fundamental to achieving expertise. Balancing technical skills with leadership and organizational abilities ensures effective management of signal processing projects, facilitating innovation and excellence in this dynamic field.

20.2 Developing Leadership in Signal Processing Strategy

Developing leadership within the sphere of signal processing strategy involves a multifaceted approach that hinges on both technical expertise and the capacity to guide and inspire teams towards innovative and effective solutions. Leadership in this domain is not solely about having a deep understanding of signal processing techniques and technologies, but also about the ability to strategically apply these skills in ways that can solve complex problems, drive research and development, and effectively steward cross-disciplinary collaboration.

One of the key elements in cultivating leadership in signal processing strategy is the emphasis on continuous learning and professional development. Signal processing is a rapidly evolving field, with new algorithms, tools, and applications emerging regularly. Therefore, a good leader must stay abreast of these advancements. This requires a commitment to ongoing education through formal courses, workshops, industry conferences, and scholarly publications. Leaders should also encourage their teams to engage in these learning opportunities to foster a culture of perpetual growth and innovation.

Another vital aspect is developing a strategic vision that aligns signal processing capabilities with broader organizational goals. A leader should have a clear understanding of the mission and objectives of their organization and be able to translate these into specific signal processing projects and initiatives. This involves identifying key areas where signal processing can add value, whether through improving operational efficiency, enhancing data analytics, or developing new products and services. A strategic vision should also account for emerging trends and potential disruptions, ensuring that the team is prepared to adapt to and leverage new opportunities as they arise.

Effective communication skills are paramount in leading signal processing teams. Complex technical concepts need to be communicated clearly and concisely to a variety of stakeholders, including non-technical team members, senior management, and clients. A strong leader must be adept at translating intricate details into accessible insights, ensuring that all parties have a shared understanding of the objectives, methodologies, and expected outcomes of signal processing initiatives. This transparency helps in building trust and securing buy-in from key stakeholders, which is crucial for the successful implementation of any strategy.

Collaboration is another cornerstone of effective leadership in signal processing. Signal processing projects typically require contributions from various disciplines, such as computer science, electrical engineering, and applied mathematics. A leader must be adept at fostering interdisciplinary cooperation, creating an environment where diverse perspectives are valued and integrated. This involves not only coordinating efforts across different teams but also managing potential conflicts and ensuring that all members feel valued and heard. Promoting a collaborative spirit helps in harnessing the collective expertise needed to tackle complex challenges in signal processing.

Moreover, a leader in signal processing should have a strong grounding in ethical considerations and be committed to responsible innovation. As signal processing techniques are often applied in areas such as surveillance, healthcare, and communications, ethical use of technology is paramount. Leaders must ensure that their strategies comply with legal standards and best practices in data privacy and security, and they should actively consider the societal impacts of their work. This involves setting up guidelines and frameworks that promote ethical decision-making and encouraging a culture where ethical considerations are routinely examined and integrated into the strategy.

Mentorship and talent development are also critical elements. Identifying and nurturing the next generation of leaders within the team ensures long-term sustainability and success. A good leader should provide mentorship, offering guidance and opportunities for emerging talent to develop their skills and take on increasing responsibilities. This involves creating clear career pathways, offering professional development opportunities, and providing regular feedback and support.

Lastly, an effective leader in signal processing strategy should be results-oriented, with a keen focus on measuring and evaluating the outcomes of their initiatives. This involves setting clear, achievable goals and establishing key performance indicators (KPIs) to monitor progress. Regular assessment and feedback loops are essential to ensure that strategies are on track and to make necessary adjustments in a timely manner. By prioritizing outcome-based management, leaders can ensure that their signal processing efforts deliver tangible benefits and drive organizational success.

In summary, developing leadership in signal processing strategy is about blending deep technical knowledge with strategic vision, effective communication, ethical responsibility, collaboration, mentorship, and a strong focus on results. Leaders who excel in these areas can not only drive advancements in signal processing but also position their organizations for sustained success in a dynamic and ever-evolving field.

20.3 Talent Management and Development in Signal Processing Teams

Talent management and development in signal processing teams are critical aspects that drive successful project outcomes and foster innovation in the field. Signal processing, an integral part of numerous applications ranging from telecommunications to biomedical engineering, hinges on the expertise, creativity, and collaboration of its workforce. Efficiently managing and developing this talent is essential for maintaining a competitive edge and achieving technological advancements.

To begin with, talent management in signal processing teams involves attracting, retaining, and nurturing skilled professionals who can contribute to various stages of signal processing including data acquisition, analysis, interpretation, and application. The recruitment process must be laser-focused, aimed at identifying candidates who not only possess robust technical skills in areas like statistical signal processing, adaptive filtering, or machine learning but also exhibit strong problem-solving abilities and adaptability. These characteristics are paramount because signal processing often involves complex and rapidly evolving challenges.

Once the right talent is recruited, the next step is to ensure their continuous professional development. This can be achieved through targeted training programs, workshops, and seminars that keep team members abreast of the latest developments and technologies in the field. Providing opportunities for further education and certification can help team members enhance their expertise and stay competitive. For example, training in advanced topics such as deep learning for signal processing or quantum signal processing can be invaluable.

Mentorship programs are a cornerstone of talent development, fostering an environment where less experienced team members can learn from veterans in the field. These programs not only facilitate knowledge transfer but also help build a cohesive team culture. By pairing novices with seasoned professionals, organizations can ensure that critical tacit knowledge and practical insights are not lost but instead perpetuated within the team.

Another key aspect of talent management is creating a collaborative and inclusive work environment. Signal processing projects often involve interdisciplinary work that spans hardware and software, theory and application, which necessitates effective teamwork. Encouraging open communication and brainstorming sessions can lead to innovative solutions and breakthroughs. Regular team-building activities and the fostering of a culture where every member feels valued and heard are equally important.

Performance assessment and feedback mechanisms are integral components of talent management. Regular evaluations, using metrics that align with both individual and team goals, help in identifying areas of strength and opportunities for improvement. Constructive feedback should be timely, specific, and geared towards professional growth. This not only aids in the personal development of team members but also enhances overall team performance.

Reward and recognition programs play a vital role in talent retention and motivation. Acknowledging and rewarding hard work, innovative ideas, and successful project completions can significantly boost morale and encourage a sense of ownership among team members. Financial incentives, promotions, public recognition, and opportunities to lead projects are various ways to acknowledge the contributions of team members.

Moreover, it is crucial to offer career advancement opportunities. Clear pathways for professional growth should be outlined, allowing individuals to envision their long-term future within the organization. This might include moving into leadership roles, transitioning to specialized technical roles, or even exploring cross-functional opportunities.

In summary, effective talent management and development in signal processing teams involve a multi-faceted approach incorporating rigorous recruitment, continuous learning, mentorship, a collaborative environment, performance feedback, rewards, and clear career progression. Focusing on these areas ensures that the team remains dynamic, innovative, and resilient in the face of evolving technological landscapes, ultimately driving forward the frontiers of signal processing.

20.4 The Role of Education and Training in Signal Processing Success

Education and training play pivotal roles in equipping individuals with the necessary knowledge and skills required for success in signal processing, an interdisciplinary field encompassing elements of electrical engineering, computer science, and applied mathematics. The rapid advancement in technology further underscores the importance of education and ongoing training as new methods, tools, and applications continue to evolve.

First and foremost, a strong foundation in relevant theoretical concepts is crucial for anyone aspiring to excel in signal processing. Education in this discipline usually begins at the undergraduate level, where students learn the fundamentals of signals and systems, discrete-time signal processing, and analog/digital signal processing. These courses typically cover essential topics such as Fourier transforms, Z-transforms, and filtering techniques. Mastery of these basics enables students to tackle more complex signal processing tasks and problems effectively.

However, theoretical understanding alone is insufficient. Practical, hands-on experience is equally important. Educational programs must incorporate laboratory work, projects, and internships to help students gain real-world experience. Lab exercises that let students work with actual signal processing hardware, such as Digital Signal Processors (DSPs) and Field Programmable Gate Arrays (FPGAs), help bridge the gap between theoretical knowledge and practical application. Projects that require implementing signal processing algorithms in software platforms like MATLAB or Python provide crucial experience in translating theory into practice.

Moving beyond undergraduate education, specialized graduate programs in signal processing offer deeper insights into advanced topics such as adaptive filtering, wavelet transforms, and machine learning applications for signal analysis. These programs often involve rigorous coursework and research components, encouraging students to contribute to the field through innovative research and development. Graduate students typically work on thesis projects that address real-world problems, further enhancing their ability to apply their knowledge effectively.

Industry-specific training is another crucial component. Signal processing professionals must stay current with the latest technologies and methodologies. Many organizations offer continuous professional development through workshops, seminars, and certifications. For example, professional bodies like the Institute of Electrical and Electronics Engineers (IEEE) provide numerous resources, including conferences, journals, and online courses, that allow practitioners to stay abreast of the latest developments. These learning opportunities are invaluable for professionals aiming to maintain a competitive edge in a rapidly changing field.

Moreover, interdisciplinary knowledge is becoming increasingly important. Signal processing applications span various industries, including telecommunications, healthcare, automotive, and entertainment. Therefore, professionals often need a working understanding of the specific industry applications of signal processing methods. For instance, in healthcare, knowledge of biomedical signal processing might be necessary to improve medical imaging techniques. In telecommunications, understanding wireless communication protocols and technologies can be crucial.

Another critical aspect is the development of soft skills. Both education and training programs must also focus on cultivating problem-solving abilities, critical thinking, and communication skills. Problem-solving and critical thinking enable professionals to devise innovative solutions to complex signal processing challenges, while strong communication skills are necessary for effectively presenting ideas and collaborating with multidisciplinary teams.

Collaboration with industry is also beneficial for educational institutions. Partnerships with tech companies and research labs allow academia to stay aligned with industry needs. These collaborations can take various forms, such as co-developed curriculum, sponsored research projects, and guest lectures by industry experts. Such synergies ensure that the education provided is relevant and up-to-date, enhancing the employability of graduates.

In conclusion, the role of education and training in achieving success in signal processing cannot be overstated. A well-rounded educational foundation combined with practical experience, industry-specific knowledge, ongoing professional development, interdisciplinary understanding, and soft skills forms the cornerstone of a successful career in this dynamic field. As technology continues to advance, the need for comprehensive education and training becomes even more crucial, ensuring that professionals are well-equipped to meet the challenges and opportunities that lie ahead.

21. Innovation in Signal Processing

The dawn of the digital age has ushered in unprecedented opportunities for advancements in various technological fields, with signal processing emerging as a cornerstone of this evolution. It is within this realm where raw data and abstract concepts converge to create meaningful, actionable information. From communication systems and medical diagnostics to multimedia applications and scientific research, signal processing has become integral to nearly every aspect of modern life.

Recent years have witnessed a surge in innovative techniques and breakthroughs that have significantly enhanced the efficiency, accuracy, and scope of signal processing. These advancements have been driven by the relentless pursuit of more powerful computational methods, increased computational power, and sophisticated algorithms capable of handling the complexity and volume of contemporary data streams.

In this chapter, we will delve into the cutting-edge innovations that are transforming the landscape of signal processing. We will explore how artificial intelligence, machine learning, quantum computing, and other emerging technologies are revolutionizing traditional methodologies, offering new paradigms and capabilities. Through case studies, expert insights, and practical applications, readers will gain a comprehensive understanding of how these innovations are being harnessed to address real-world challenges and open new frontiers.

Join us as we embark on an exploration of the future of signal processing—a future filled with promise, potential, and profound impact on the way we interact with and interpret the world around us.

21.1 Leveraging Innovation for Signal Processing Success

In the ever-evolving landscape of signal processing, innovation stands as the linchpin for achieving and maintaining success. As the field of signal processing continuously intersects with other rapidly advancing sectors such as telecommunications, medical technology, and artificial intelligence, leveraging innovation becomes not only advantageous but also essential. Innovation drives both the theoretical advancements and the practical implementations that define signal processing, fundamentally reshaping how data is interpreted, manipulated, and utilized.

At the core of innovation in signal processing are the breakthroughs in algorithmic development. Traditional methods, such as Fourier Transform and Wavelet Transform, have long been the bedrock of the field. However, with the integration of machine learning algorithms and neural networks, the scope and efficiency of signal processing have reached unprecedented levels. Machine learning algorithms, particularly deep learning models, excel in identifying patterns and features within data that are not easily discernible through classical techniques. These models can automatically learn from large datasets, adapt to new types of signals, and provide more accurate and comprehensive analysis. Consequently, the fusion of deep learning with signal processing has led to enhancements in areas such as image recognition, speech processing, and even real-time data analysis.

Moreover, the hardware on which signal processing is performed has seen significant leaps in innovation. With the advent of specialized processing units like Graphics Processing Units (GPUs) and Tensor Processing Units (TPUs), the ability to perform complex computations in parallel has increased manifold. These advanced processors allow for the execution of sophisticated algorithms in real-time, fostering applications that demand high-speed and high-accuracy signal interpretation. For instance, the use of Field-Programmable Gate Arrays (FPGAs) enables customizable, high-performance computing solutions tailored specifically for signal processing tasks. This synergy between cutting-edge hardware and algorithmic advancements ensures that signal processing applications are not only faster but also more energy-efficient and adaptable.

Another vector of innovation in signal processing is the heightened focus on interdisciplinary collaboration. The integration of different fields such as neuroscience, bioinformatics, and environmental science with signal processing has yielded revolutionary applications. For instance, in the realm of medical technology, innovative signal processing techniques are pivotal in the development of sophisticated diagnostic tools and therapies. Techniques like magnetoencephalography (MEG) and functional Magnetic Resonance Imaging (fMRI) rely on advanced signal processing to map brain activity with high precision, thus providing invaluable insights for neuroscience research and clinical diagnosis. Similarly, in environmental science, signal processing helps in analyzing satellite imagery and sensor data to monitor and mitigate the effects of climate change.

Furthermore, the rise of Internet of Things (IoT) technology has propelled the necessity for innovative signal processing solutions. IoT devices generate massive amounts of data that require real-time processing and analysis. Adaptive signal processing techniques, leveraging adaptive filters and machine learning, ensure efficient handling of this data influx. For example, smart city initiatives use such techniques to manage and optimize resources like water and electricity, improve traffic management, and enhance public safety through real-time monitoring systems.

Innovation is also redefining the paradigms of data security and privacy in signal processing. As signals often contain sensitive and personal data, advanced encryption methods and secure processing techniques are crucial to protect this information. Techniques like homomorphic encryption enable computations to be performed on encrypted data without needing to decrypt it first, thereby ensuring privacy while maintaining the integrity and utility of the data.

In summary, leveraging innovation in signal processing is pivotal for success in the contemporary technological ecosystem. The convergence of advanced algorithms, high-performance hardware, interdisciplinary applications, and secure processing paradigms defines the frontier of this field. Continued innovation will undoubtedly lead to smarter, faster, and more reliable signal processing solutions that will permeate every aspect of modern life, from healthcare to environmental monitoring, to everyday consumer technologies. In this relentless pursuit of innovation, the signal processing community advances not only its own boundaries but also pushes the envelope of technological progress itself.

21.2 Innovative Approaches to Signal Processing

In the evolving realm of signal processing, innovative approaches are continuously being developed to address the increasingly complex challenges posed by advancements in technology. These novel methodologies often strive to enhance the efficiency, accuracy, and applicability of signal processing techniques across various domains, ranging from telecommunications to biomedical engineering. One of the foremost areas of innovation in this field is the application of machine learning and artificial intelligence.

Machine learning algorithms have revolutionized signal processing by enabling the development of adaptable and robust models that can manage and interpret vast amounts of data. Traditional signal processing techniques hinge on deterministic approaches and mathematical formulations to isolate and analyze signals. However, machine learning introduces the ability to discern patterns and relationships in data without explicit programming, thereby automating the classification, filtering, and prediction processes. For instance, deep learning networks, particularly convolutional neural networks (CNNs), excel in tasks such as image and speech recognition due to their proficiency in processing high-dimensional inputs and detecting intricate features within the data.

Another groundbreaking approach is the utilization of compressed sensing, which capitalizes on the sparsity inherent in many natural signals to significantly reduce the number of samples required for accurate signal reconstruction. This technique challenges the classical Nyquist-Shannon sampling theorem by asserting that it is possible to reconstruct signals from far fewer samples than traditionally deemed necessary, provided that these signals are sparse in some domain, such as the frequency or time-frequency domain. Compressed sensing has profound implications in various fields, including medical imaging, where it can dramatically decrease the duration and radiation exposure of MRI and CT scans without compromising image quality.

In parallel, adaptive signal processing represents an equally impactful innovative approach. Adaptive algorithms dynamically adjust their parameters in real-time to best capture and process changing signal characteristics. This is particularly useful in scenarios involving non-stationary signals, where the statistical properties of the signal vary with time. The application of adaptive filtering in noise cancellation, echo suppression, and wireless communications has showcased its potential to maintain signal quality under fluctuating conditions, thus enhancing the robustness and reliability of these systems.

Nonlinear signal processing is another frontier that addresses the limitations of linear methods when dealing with real-world signals that exhibit nonlinear behaviors. Traditional linear techniques often fall short in accurately capturing such complexities, leading to the development of nonlinear approaches that can better model and predict the behavior of these signals. Techniques such as chaos theory and fractal analysis are employed to unravel the intrinsic characteristics of nonlinear systems, which are prevalent in areas like finance, geophysics, and biological systems.

Furthermore, the integration of quantum computing with signal processing has opened unprecedented avenues for innovation. Quantum algorithms, leveraging principles such as superposition and entanglement, promise to solve certain types of signal processing problems exponentially faster than classical algorithms. For example, quantum Fourier transform and quantum principal component analysis hold the potential to revolutionize frequency analysis and data compression. While the practical implementation of quantum signal processing is still in its nascent stages, ongoing research indicates immense promise for future breakthroughs.

In addition to these methodologies, hybrid approaches that combine multiple novel techniques are gaining traction. For instance, combining machine learning with compressed sensing can lead to systems that are both highly efficient in data acquisition and powerful in pattern recognition. Similarly, the fusion of adaptive and nonlinear methods can yield adaptive nonlinear filters that offer tailored solutions to complex, time-variant signal processing challenges.

In summary, the landscape of signal processing is being reshaped by a myriad of innovative approaches that push the boundaries of traditional methods. By integrating advancements from machine learning, compressed sensing, adaptive algorithms, nonlinear analysis, and quantum computing, these novel techniques not only enhance the capabilities of existing applications but also pave the way for new possibilities in processing and interpreting signals. As technology continues to advance, the synergy between these innovative approaches will likely lead to even more sophisticated and effective signal processing solutions.

21.3 The Role of R&D in Advancing Signal Processing Practices

Research and Development (R&D) has always been a cornerstone in the advancement of signal processing practices. Signal processing, integral to a broad range of applications such as telecommunications, medical imaging, audio and video encoding, and radar systems, has been significantly enhanced through continuous R&D endeavors. The improvements in methodologies, algorithms, and technologies within signal processing owe much to structured research frameworks and developmental projects.

One of the primary roles of R&D in advancing signal processing practices involves the discovery and optimization of algorithms. Algorithms are the mathematical calculations and procedures that allow signal processors to filter, compress, and transform signals into usable data. Advances in R&D have led to the creation of more efficient algorithms that can process signals faster and more accurately. For example, the development of Fast Fourier Transform (FFT) algorithms revolutionized the way frequency analysis is conducted, making it possible to perform rapid and efficient transformations from time domain to frequency domain.

Moreover, R&D in signal processing has facilitated the design and creation of adaptive algorithms that can dynamically adjust to changing signals or conditions. This is particularly important in environments where signal characteristics can vary over time, such as wireless communications. Adaptive algorithms have enhanced error correction, signal decoding, and extraction, making communication systems more reliable and resilient.

Signal processing hardware has also seen significant advancements due to R&D efforts. Modern signal processing tasks require highly specialized hardware to meet the demands of high-speed processing and real-time applications. Research in this area has led to the development of Digital Signal Processors (DSPs), Field-Programmable Gate Arrays (FPGAs), and Application-Specific Integrated Circuits (ASICs). These hardware innovations provide the necessary computational power and efficiency required for sophisticated signal processing tasks.

In addition to hardware, software development in signal processing has been transformative. Software-defined radio (SDR), for instance, is an outcome of extensive R&D in signal processing software. SDRs allow for the processing of signals using software rather than hardware, providing flexibility and the ability to upgrade systems through software updates. This innovation means that new signal processing techniques can be implemented without the need for new hardware, significantly reducing costs and deployment times.

R&D has also played a pivotal role in the interdisciplinary integration within signal processing domains. Areas such as machine learning and artificial intelligence are increasingly being incorporated into signal processing practices, thanks to research initiatives. Machine learning algorithms are now used to improve signal detection, classification, and prediction tasks. For example, neural networks can be trained to identify patterns and anomalies in signals, which is useful in applications ranging from medical diagnostics to cybersecurity.

Furthermore, R&D in signal processing has contributed to the development of new standards and protocols. These standar.ds ensure compatibility and interoperability among different systems and devices, which is crucial in a globally connected world. The development of compression standards such as JPEG for image compression and MP3 for audio compression are direct outcomes of significant R&D efforts in signal processing. These standards have not only improved the efficiency of data storage and transmission but have also enabled the widespread adoption of digital media technologies.

In the medical field, advances in signal processing through R&D have led to better diagnostic tools and therapeutic devices. Techniques such as Magnetic Resonance Imaging (MRI) and Electroencephalography (EEG) rely heavily on sophisticated signal processing algorithms to interpret the captured signals. Research in this area has led to improved image quality, faster processing times, and more accurate diagnostic capabilities, ultimately enhancing patient care and outcomes.

Lastly, the role of R&D in addressing emerging challenges in signal processing cannot be understated. As the volume and complexity of data continue to grow, R&D is essential in developing new solutions to handle big data and ensure efficient and reliable signal processing. This includes addressing issues such as signal noise, interference, and the necessity for real-time processing in increasingly complex environments.

In conclusion, the role of R&D in advancing signal processing practices is multifaceted, encompassing the development of sophisticated algorithms, specialized hardware, and innovative software solutions. These advancements are crucial for improving the efficiency, accuracy, and reliability of signal processing across various domains. Continued investment in R&D is essential to address emerging challenges and to harness new opportunities in the ever-evolving field of signal processing.

21.4 Case Studies of Innovation in Signal Processing

In examining the landscape of signal processing innovations, several exemplary case studies stand out for their transformative impact on both technology and society. Signal processing—an essential component in telecommunications, data compression, speech and image recognition, and various other fields—continuously evolves through pioneering research and real-world applications. This section delves into specific instances where innovative approaches in signal processing have led to significant advancements, offering insights into the mechanisms and achievements in this critical area.

One remarkable case study is the development of adaptive filter algorithms, which revolutionized the way signals are processed in noisy environments. Adaptive filters are designed to automatically adjust their parameters to minimize the error between the processed signal and a desired signal. These algorithms have been instrumental in applications ranging from noise cancellation in headphones to echo cancellation in telecommunication systems. An example is the work conducted by Widrow and Stearns, whose Least Mean Squares (LMS) algorithm laid the groundwork for many subsequent adaptive filtering techniques. This innovation has greatly improved the quality of audio and communication systems, making it possible to extract clear signals from a cacophonous backdrop.

Another significant case pertains to the advancements in medical imaging, particularly in Magnetic Resonance Imaging (MRI). Traditional MRI techniques faced limitations in terms of resolution and speed, which impeded diagnostic capabilities. However, the introduction of compressed sensing algorithms marked a breakthrough. Developed by Emmanuel Candès and Terence Tao, compressed sensing leverages the sparsity of signals to reconstruct high-quality images from fewer samples than traditionally required. This approach has dramatically accelerated MRI scan times while maintaining, or even improving, image quality. It has not only enhanced diagnostic precision but also reduced patient discomfort and operational costs in medical facilities.

The realm of digital communication has also benefitted immensely from innovative signal processing techniques. Orthogonal Frequency Division Multiplexing (OFDM) is a key example. OFDM, which divides a signal into multiple smaller sub-signals that are transmitted simultaneously at different frequencies, mitigates the effects of multipath interference and improves bandwidth efficiency. Initially applied in military and professional communication systems, OFDM was eventually adopted into various commercial technologies, including Wi-Fi, LTE, and digital television broadcasting. The transition to OFDM-based systems has facilitated higher data rates and more reliable communication, underpinning the modern digital communication infrastructure.

Speech and audio processing have similarly seen phenomenal advancements. One illustrative case is the evolution of Automatic Speech Recognition (ASR) systems. Early ASR systems, which relied on template matching techniques, were limited in vocabulary and speaker-specific. The introduction of Hidden Markov Models (HMMs) revolutionized the field by enabling more flexible and accurate recognition of continuous speech. Innovations continued with the advent of deep learning-based techniques, where convolutional and recurrent neural networks (CNNs and RNNs) began to outperform traditional models. Companies like Google and Amazon have integrated these advanced ASR systems into their products, such as virtual assistants and smart home devices, significantly enhancing user interaction capabilities.

In the field of image processing, breakthrough innovations such as wavelet transforms have set new standards. Unlike traditional Fourier transforms, which only capture frequency information, wavelet transforms provide both time and frequency localization, offering a more nuanced analysis of images and signals. This technology has been pivotal in applications ranging from image compression (e.g., JPEG 2000 standard) to medical diagnostics, where detailed image analysis is paramount.

Financial market analysis has also benefitted from signal processing innovations, particularly in algorithmic trading systems. Techniques such as Fast Fourier Transforms (FFT) and Kalman Filters have been employed to analyze and predict market trends. These tools allow for the real-time processing of vast datasets, enhancing the ability to make quick and informed trading decisions. As a result, they have become integral to the operations of hedge funds and financial institutions.

In conclusion, these case studies highlight the diverse applications and profound impact of innovative signal processing techniques. From enhancing communication and medical diagnostics to improving user interactions and financial operations, the advancements in signal processing continue to play a pivotal role in the development and optimization of contemporary technologies. Each case underscores the ongoing need for research and development in this field, as future innovations promise to address emerging challenges and opportunities in an increasingly data-driven world.

Cybellium Ltd – Understanding Signal Processing

www.ingramcontent.com/pod-product-compliance
Lightning Source LLC
La Vergne TN
LVHW090311210926
848480LV00001B/209
* 9 7 8 1 8 3 6 7 9 1 0 8 9 *